Design for Manufacture: Principles and Practices

Henry W Stoll

Professor Emeritus of Mechanical Engineering
Northwestern University

ISBN: 9798621204259

Copyright © 2020 by Henry W. Stoll. All rights reserved.

Available: Amazon.com

Preface

The importance of design for manufacture (DFM) is now recognized worldwide. This is a book about the principles that underly DFM and the practices that facilitate its implementation. It presents a systematic and holistic DFM approach that is easy to understand, remember, and apply. Writing a book such as this has been in my thoughts since a design manager attending one of my DFM workshops asked me if there was a book on DFM that covered material similar to what was being presented in the workshop. What he wanted was a book that captures the essence of DFM in terms that practicing design and manufacturing professionals will immediately find relevant and useful. My hope is that this book helps fulfill that need.

Because it emphasizes the philosophical side of DFM rather than the technical side, the book should be of interest to all levels of manufacturing enterprise management as well to design and manufacturing professionals who desire to continually improve. The book should also be suitable as a supplementary text in business school courses on product development, and for undergraduate and graduate engineering courses in design and manufacturing.

Henry W. Stoll

Contents

Preface		*iii*
Chapter 1	Design for Manufacture and the Manufacturing Enterprise	1
Chapter 2	Good Design	13
Chapter 3	The Design-Manufacture Interface	24
Chapter 4	Design Descriptions	52
Chapter 5	The DFM Approach	74
Chapter 6	Simplicity: The Holy Grail of DFM	92
Chapter 7	Undesirable Interactions: The Hidden Menace	102
Chapter 8	Standardization: DFM's Secret Weapon	114
Chapter 9	Part Count Reduction: The First Law of DFM	133
Chapter 10	Design for Assembly: The Second Law of DFM	145
Chapter 11	Piece-Part Cost Reduction: The Third Law of DFM	155
Chapter 12	The Geometric layout Improvement Method	167
Chapter 13	Robust Design: Combating Variation and Change	176
Chapter 14	Coordinated Design: Integrating Product and Process	187
Chapter 15	Ten DFM Success Factors	210
Chapter 16	Producibility Checklist	225
Chapter 17	The Essence of DFM	232
References		235
About the Author		237

Design for Manufacture:
Principles and Practices

Chapter 1
Design for Manufacture and the Manufacturing Enterprise

This book is about Design for Manufacture, or DFM for short. DFM is the general engineering practice of designing products in such a way that they are easy to manufacture, assemble, and support. The aim of the book is to present DFM in terms of the fundamental principles of good design that underly it and to use these principles as the basis for a holistic and systematic DFM approach.

Although manufacturing has always been an important design consideration, it rose to critical importance in the early 1980's when the computer and the design and manufacturing technologies it enabled rapidly revolutionized the way design and manufacturing is done. DFM as a formal discipline was born out of this revolution. In this chapter, the concept of DFM is placed into the broader context of the manufacturing enterprise. The relationship between DFM and the manufacturing enterprise is a constant consideration that affects every aspect of DFM and its implementation.

The Importance of Design

The importance of design should never be underestimated or taken for granted. The business goals of most manufacturing enterprises are two-fold: (1) make as good a product as possible, in as short a time as possible, and for as little cost as possible; and (2) sell as many as possible, as fast as possible, for as much as possible. Design, more than anything else, determines the firm's ability to successfully achieve these goals. It determines the desirability, quality, and reliability of the product, the development time required to bring it to market, and most importantly, the cost and complexity involved in its production. Moreover, design determines the firm's long-term competitive position and reputation. From a DFM perspective, design is particularly important for two crucial reasons:

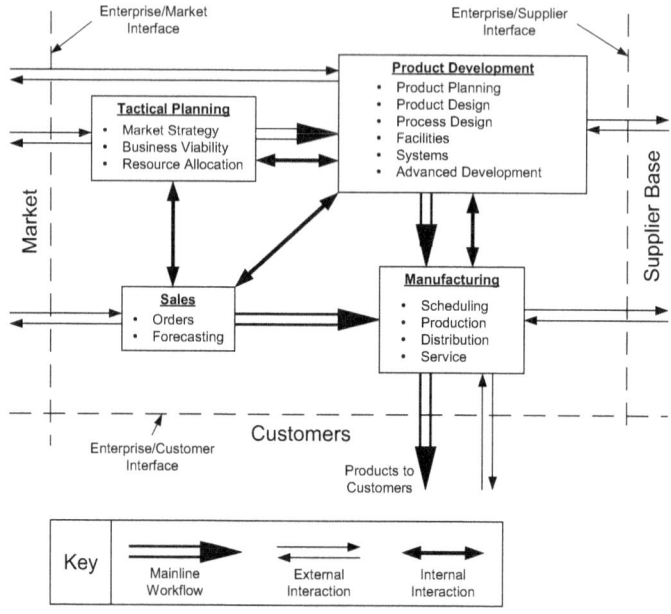

Figure 1.1 Workflow model of the manufacturing enterprise.

1. **Design is a Strategic Activity:** Whether by intention or by default, design establishes the firm's manufacturing strategy.

2. **Design is the First Manufacturing Step:** Before a product can be manufactured and sold, it must be designed. Because design comes first, it directly dictates up to 80% or more of manufacturing decisions.

The Manufacturing Enterprise

Consider the simple model of a manufacturing enterprise shown in Fig. 1.1 (Ettlie and Stoll, 1990, Chapter 7). In this model, the "basic actions" of a manufacturing enterprise are divided into four main groupings: (1) decide what customers want, (2) set up the factory to make it, (3) produce it, and (4) sell and support it. Tracing the "main-line" generic workflow that connects these activities together yields the two major cycles on which most manufacturing enterprises operate: (1) the order-to-delivery cycle, which involves the receipt of orders and production of existing products, and (2) the new product introduction cycle, which involves planning and introducing new products and existing product improvements and upgrades into production.

The Order-to-Delivery Cycle: The order-to-delivery cycle is the operations side of the business; it generates the day-to-day cash flow for the firm by producing and distributing products, assuring product quality, and providing the interface between the company and its customers in the form of sales and service. Major activities typically include order fulfillment, production, and service. The order fulfillment activity brings in product orders (sales) and forecasts production requirements, production converts sales orders into products, and service provides product maintenance, repair, and disposal support to customers.

New Product Introduction Cycle: New products may be novel products that the firm is introducing for the first time, or they may be new versions, redesigns or upgrades of existing products. The new product introduction cycle involves two major activities: (1) tactical planning and (2) product development. Tactical planning determines the "what" and "when" for product design and development and quantifies the needs that guide design and manufacturing. Product development translates the "what" into "how" by defining the new product designs to be introduced into production and providing the manufacturing plan, physical hardware, systems design, information, training, and guidance to make new processes, tooling, and equipment work in production.

DFM seeks to simplify, smooth, and shorten these two cycles. Customer satisfaction is improved because improved product quality and short cycles gives customers what they want when they want it. Manufacturing cost and time are reduced because quality risk, waste and non-value-added activity are eliminated across the entire manufacturing enterprise. The result is improved enterprise profitability. Market share is increased and the ability to respond quickly and flexibly to changing market conditions with minimum cost impact is improved.

Manufacturing System

To design and manufacture products, the manufacturing enterprise typically employs a manufacturing system that is set up to convert starting material into finished products that are marketed and sold for a profit. The manufacturing system includes all activities required to bring a product to market including product conception, design and development, marketing, sales, purchasing, manufacturing, assembly, distribution, and lifecycle support. Embedded within the manufacturing system are many distinct processes and activities that, individually and collectively, affect product

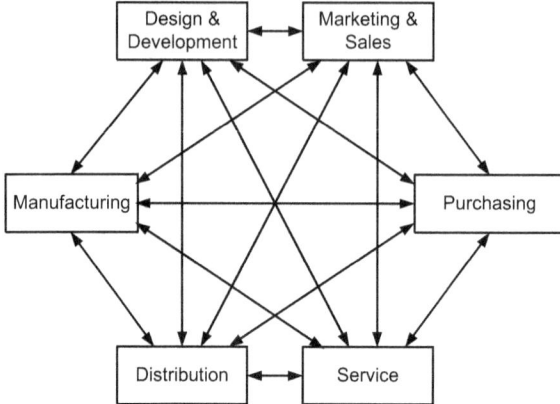

Figure 1.2 Interplay between components of the manufacturing system.

quality and cost, as well as the time and effort required to manufacture the product and the time and effort required to design and introduce a new version of the product. Interplay between these many processes and activities is complex, and decisions made concerning one aspect have ramifications that extend to the others (see Fig. 1.2).

In its broadest sense, design for manufacture seeks to comprehend the complex interplay between the components of the manufacturing system and use the understanding and insight gained to better design the products and equipment manufactured and sold by the manufacturing enterprise. More specifically, DFM seeks to:

1. Understand how the process by which a product is designed interacts with the manufacturing system and use this understanding to design better quality products that can be produced for lower cost and brought to market more quickly.

2. Understand how the physical design of the product itself interacts with the manufacturing system and use this understanding to define product design alternatives that facilitate "global" optimization of the product and process as a coordinated system.

Design Situations

Design projects can be categorized depending on the design and manufacturing situation that is involved (see Fig. 1.3). A *new design* is one that involves significant change from existing models or versions of the product. Significant change in this case implies a major deviation from current practice such as a new working principle or a new and different

		Method of Manufacture	
		New	Existing
Design	New	Situation #1	Situation #2
	Existing	Situation #3	Situation #4

Figure 1.3 Design and manufacturing constraint matrix.

technology. An *existing design*, on the other hand, is a current product or design that is redesigned to reduce manufacturing cost and/or to improve performance and correct quality issues. This includes new models or variants designed to meet niche or new customer or market needs.

With regard to manufacturing, a new design may be produced using either tried and true existing methods or by using completely new methods. A *new manufacturing* method is one that has not been used previously by the company to produce the design. Conversely, an *existing manufacturing method* is one that is currently being used in production. Creating a design that is to be produced using a totally new to the company manufacturing method is quite different from one that is to be made using existing, well-understood methods and processes. Switching from a sheet metal housing to a molded plastic housing, for example, would be a new manufacturing method for a company that has traditionally made sheet metal housings and that has a plant in which production is mostly performed by stamping and sheet metal forming equipment.

With these concepts in mind, four design situations can be envisioned using the matrix in Fig. 1.3. These situations can be thought of as forming the corners of a field in which most design projects are located.

1. A new design and a new manufacturing method.
2. A new design and an existing manufacturing method.
3. An existing design and a new manufacturing method.
4. An exiting design and an existing manufacturing method.

Most design projects fall somewhere within the bounds formed by these extremes. Each new design project offers different DFM opportunities and challenges. Examination of these extreme situations provides insight into the variety of design approaches and potential benefits that are possible.

Design Situation #1

This design situation is the best possible for DFM because neither the design nor the manufacturing method is constrained by past practices or fixed investments. Unfortunately, because both the design and the process are new, design and manufacturing knowledge is probably the least complete. Also, the budget of time and money may be short. The challenge, therefore, is to take the time, up front, to correctly understand the problem of design fully so that the most optimal early design decisions can be made. The key is to let the design axioms guide every design decision (see Chapters 2, 6 and 7). By strictly adhering to these axioms and coordinating the product and process (see Chapter 14), the ability to evolve the design as design knowledge becomes more complete should be relatively straight forward with a minimum of ripple effect.

Design Situation #2

This situation commonly occurs in vertically integrated companies that produce a range of products using well-established manufacturing methods. The DFM opportunity offered by this situation is to create new designs that readily map onto the existing production facility with the least disruption and cost. The challenge in doing this is to correctly understand and characterize the manufacturing needs and constraints that exist and then to package this information in a form that is readily usable by the design team. One possibility is to develop a *designer toolkit* that provides design rules and guidelines, physical examples and models, computerized design aids, and other specific information about the manufacturing facility in an easily used form. Such an approach is an example of process-driven design. Process-driven design is a design strategy in which the method of manufacture is specified "up front" as a design requirement.

Design Situation #3

This situation is typical in companies that have been in business for relatively long periods of time. It is not unusual for companies such as this to manufacture and sell literally hundreds to hundreds of thousands of "mature" products and product models. Mature products have usually been (1) designed and developed one at a time over many years, (2) are part of a large portfolio of products that has been expanded through evolutionary and chronological developments, and (3) do not embody much compatibility, commonality, standardization, or modularity.

A significant opportunity in this design situation is to leverage the concepts of standardization (see Chapter 8) and coordinated design (see

Chapter 14) together with flexible automation and advanced manufacturing technology to achieve the following goals:

- Family of products derived from one "maximum" design.
- All models on one production line.
- Resetting for type change in minutes.
- New variants quickly and easily designed and introduced.
- New technology, materials, and processes easily introduced.

Achieving these goals usually requires a total redesign of all existing products. The opportunity is to create a coordinated and unified product family that makes all subsequent redesigns within a given product family Situation #2 type projects. The challenge is to redesign the existing products in such a way that these goals can be achieved while, at the same time, adhering to defined parameters regarding functionality, working principle, outward appearance, and so forth. This is especially true for industrial products such as circuit breakers, valves, and switches that are often used as components in larger systems and are also often subject to code and other regulatory approvals. Designs in which the design and the production method are coordinated is one way of solving this problem (see Chapter 14).

In some mature products, it may be possible to replace costly or failure prone components with new technology, materials, or processes. For example, a complex weldment involving the joining of several component parts could be replaced with a thin wall, light-metal sand casting. New layered manufacturing technologies have reduced tooling cost to the point where casting is a viable alternative, even in applications involving low production quantities. The identification and re-engineering of conversion parts is an important aspect of Design Situation #3 for reducing cost and improving quality, especially in products where a total redesign is not economically practical.

Design Situation #4

This situation arises when problems develop shortly after new product launch or, what is more often the case, when initial manufacturing costs are higher than expected. In these cases, resolving this situation can be particularly difficult because of the constraints imposed by the existing design, tooling and equipment investments, and manufacturing method. Frequently in these situations, only minor design changes are allowable, and resources are usually limited. To make matters worse, any change that is made can necessitate other changes that sooner or later result in suboptimal design and undesirable compromise. For this and many other reasons, Design Situation #4 should be avoided by using a DFM approach.

At the same time, experience with methods such as the "geometric layout improvement method" discussed in Chapter 12 and the Boothroyd-Dewhurst design for assembly method has shown that the geometric layout of existing designs can often be improved significantly with only minor changes. When this is the case, and it frequently is, the DFM approach can be used to cost reduce the design, eliminate manufacturing quality issues, and greatly simplify the assembly process.

Design Strategy

The challenge of "good design" is to identify the best design out of the universe of possible designs. The value of a clearly articulated design strategy is that it helps to eliminate and narrow choices, makes the best choices more obvious, and constrains choices that would otherwise be arbitrary. From a DFM perspective, three possible approaches for developing an effective design strategy are possible:

1. Use the design axioms as illustrated by Example 2.2 in Chapter 2.
2. Use a coordinated product and process strategy (see Chapter 14).
3. Develop a business strategy that guides the design.

As indicated, the first two of these strategies are discussed elsewhere in the book. The third strategy recognizes that how a product is designed depends in large measure on the way the product is to be manufactured and brought to market. A wisely conceived business strategy provides needed guidance and direction to the design process. To appreciate how business strategy can be used to drive design, consider two companies, Company "A," and Company "B," that manufacture and sell competing products in the same market.

Company A: The marketing strategy of this company is to sell several different product models, differentiated based on cost and performance, through large retail outlet chains and other "big box" mass retailers. To support this marketing strategy, a manufacturing strategy is developed in which all models are mass produced on one assembly line and delivered in scheduled shipments to retail outlets using an optimized distribution system. With the marketing and manufacturing strategies thus determined, the design strategy becomes obvious: design the product so that all models can be mass produced interchangeably on one production line and packaged such that specified quantities of each model can be easily sorted and palletized for scheduled shipment to specific retailers.

Company B: The marketing strategy of this company, on the other hand, is to sell its product directly to the end user over the internet and in this way allow the customer to customize the product in any way they want via their internet order. This marketing strategy calls for a manufacturing strategy in which the customized product is completed as part of the order-to-delivery process. The resulting design strategy calls for a master design that is completely customizable as a last step in the production process. This step may be performed either at the last station on the production line or in a small production facility located at each remote regional distribution site.

Because of the different strategies used, design choices and decisions for each company are quite different. What constitutes a good design for manufacture for Company "A" is clearly different from that for company "B." As this example illustrates, design strategy can totally change both the design and the way it it is manufactured.

Beneficial Effects Produced by DFM

The tangible benefits of DFM are many and varied. In many cases, specific benefits depend on the design situation. In general, manufacturing companies that have been using DFM on a regular and consistent basis report the following benefits of DFM.

- Significant reduction in manufacturing cost.
- Significant improvement in manufactured quality.
- Easier and faster product development and launch.
- Significant reductions in warranty claims and product recalls.
- Impressive reductions in assembly time, errors, and rework.
- Dramatic reduction in tooling, fixturing, and production complexity.
- Greatly simplified material handling and logistics.
- Improved cooperation between all stakeholders.
- Increased opportunities for automation.
- Substantial reductions of engineering change notices.

These benefits result from a combination of the many ways that DFM works to improve the design. Some of the most effective of these include:

DFM is Good Design: In many respects, this book is as much about good design as it is about design for manufacture. The concept of good design can be defined as the timely design of functionally and aesthetically appealing products that have inherent high quality, low cost, and ease of manufacture. *Timely* implies the ability to quickly create products to meet new market needs. *Functionally and aesthetically appealing* implies a design that meets well defined and clearly understood customer needs in a way that delights the customer and ensures long-term, sustainable customer satisfaction. *Inherent* implies that high quality, low cost, and ease of manufacture are ingrained properties of the design that have been "designed-in". Each of these properties have special meaning. In good design, the term *quality* is used in its broadest sense to mean a product that provides the performance and features the customer wants, unswervingly conforms to design intent, is robust against hard to control variation and change, is safe, and is designed to have the reliability, availability, maintainability, and dependability (RAM-D) required. Similarly, the term *cost* refers to total lifecycle cost and includes direct cost, tooling and investment cost, system cost, development cost, and cost of time. Finally, *ease of manufacture* implies that the method of manufacture and assembly has been considered from the inception of the design. Design for manufacture, in its most general sense, is a collection of design principles, strategies, and methodologies that help facilitate and ensure good design.

DFM Simplifies and Integrates Product and Process: DFM focuses on downstream processes to help ensure products that are easy to manufacture, assemble, repair, service, recycle, and improve over time. This produces an extensive reduction in product and manufacturing process complexity as well as substantial improvements in the coordination between the product design and the production process.

DFM Improves Manufacturing Productivity: Increased design simplicity and product and process coordination results in reduced cycle times and other processing times involved in activities such as setup and material handling, as well as reduced quality risk that causes line shutdowns and product rework. The result is increased productivity.

DFM Improves the Quality of Early Design Decisions: Early design decisions affect lifecycle cost far more than years of manufacturing improvements made after concept decision and detail design (see Fig. 1.4). The elimination of a part, for example, or a machining direction or a separate fastener, by design can result in the elimination of workstations, operations, fixturing, and quality risks, together with the direct costs that accompany them, for the life of the product. No amount of optimization of speeds and feeds or work simplification or advanced manufacturing technology can match the benefits of a product designed, from the outset, for manufacture. By ensuring that the voices of "downstream" customers are heard in the early stages of the design process, DFM helps make manufacturing related design information available when its needed.

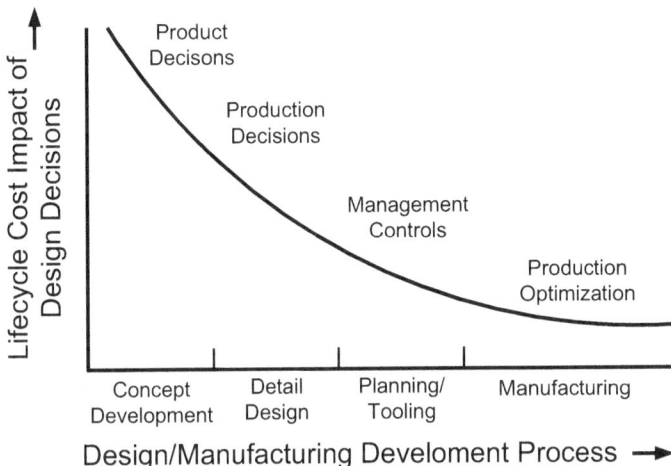

Figure 1.4 Cost impact of design decisions.

DFM Produces Self-Compounding and Synergistic Benefits: DFM generates a blend of design improvements that tend to self-compound and synergistically combine in ways that produce overall productivity improvements that are much greater than the sum of individual effects. This in turn further reduces cost because less resources are needed to produce more product in less time. The positive self-compounding and synergistic effect of DFM explains why its benefits are so dramatic and wide-ranging.

DFM Reduces Engineering Change: By spending time upfront to develop more complete design information, fewer engineering changes are required on the way to production and sales. This means a shorter design cycle with less design compromise and chaos and, hence, a product that not only launches on schedule, but also more closely represents original design intent thus further compounding the firm's ability to make as good a product as possible, in as short a time as possible, and for as little cost as possible.

DFM Serves as a Common Language: Finally, the basic tenets of DFM serve as the foundation for a system of beliefs and understanding that creates a common language among all stakeholders in the design. This common language not only facilitates effective DFM, it also enhances and improves communication across the entire manufacturing system because everyone has a shared understanding. Experience has consistently shown that good communication between all stakeholders in the design is the single most important determinant in DFM success (see Chapter 15).

About the Book

Much of the material in this book is taken from manufacturing course notes and DFM workshops that I have taught for many years and that I have published in various forms elsewhere. Like the process of design itself, the book flows from the abstract to the concrete. This and the next three chapters (Chapters 1-4) provide the context for DFM. Chapter 5 presents the DFM approach and sets the stage for all that follows. Chapters 6 and 7 present fundamental principles of design that underly all aspects of the DFM approach. In Chapters 8 through 13, these principles are applied across the manufacturing system in a variety of ways. Finally, Chapters 14 through 17 tie everything together and chart a path for DFM success. Most books on design for manufacture focus on cost reduction. This book is different. It promotes the view that DFM is a philosophy of design that guides and informs all aspects of the design realization process.

Summary of Key Concepts

- ➤ The manufacturing system is complex and tightly coupled. Successful DFM must navigate this complexity to succeed.
- ➤ DFM depends on the design situation and the design strategy used.
- ➤ The benefits produced by DFM are often magnified by the synergistic and self-compounding productivity improvements that DFM enables.
- ➤ DFM is the common language of design and manufacturing.

Chapter 2
Good Design

Good design is an ideal toward which the engineer, the design team, and the manufacturing enterprise strive but seldom achieve perfectly. The concept of good design underlies every aspect of design for manufacture; it is the bedrock that supports the entire edifice of DFM. In this chapter, we seek to understand the concept of good design in terms of what it is, how it might be implemented, and how its effect on the success of the manufacturing enterprise may possibly be qualitatively measured.

What is Good Design?

Like beauty, good design is in the eye of the beholder. From the perspective of design for manufacture, a good design is one that not only fulfills its purpose but is also one that is easy to manufacture and support over its lifecycle. In this context, a *good design* can be defined as one that accords with the *rules of good design*:

Rule of Needs. *Customer needs are to be comprehensively understood and then imaginatively satisfied by the design.* Designs that adhere to this rule invariably outperform and outsell competitor products while costing less to produce and support. The needs of all customers must be understood and satisfied, including considerations such as environmental impact, and material and purchased part availability. In general, customer needs are hierarchically ordered in priority, with end user needs and manufacturing needs at the top.

Rule of Clarity. *All aspects of the design should be predictable and unambiguous.* "Clarity" is the quality of being coherent and intelligible. This means that the design and the manufacturing system that produces it function and behave in understandable and predictable ways, both with use and over time. Performance and the effect of hard-to-control variation are predictable and unambiguous. There are no "surprises", unexplained behaviors, mysterious uncertainties, manufacturing glitches, or hard to resolve quality issues. When and if engineering change is necessary, changes can be made without undesirable ripple effects and subsequent loss of optimality. All design information including instructions, drawings, and specifications are intelligible, unambiguous, logical, and consistent.

***Rule of Simplicity*.** *The design should be "not complex", "easily understood", and "easily done".* We tend to think of simple things as being beautiful. They are beautiful because they are more economical, with less waste, less superfluous clutter. Design for manufacture can, on a very fundamental level, be thought of as the ultimate pursuit of simplicity. In a simple design, the number of parts, processes, process steps, tools, adjustments, and so forth are held to an absolute minimum. Parts are shaped so that they are easy to make and assemble. Working principles are chosen based on their ability to be efficient, that is, to be productive without waste. Design simplicity improves reliability, eliminates quality risk, and reduces total cost and time. It reduces waste and hard-to-control variation, the two sworn enemies of manufacturing and lifecycle support.

***Rule of Safety*.** *The design should function reliably and safely without harming or endangering humans or the environment.* In a safe design, all types and modes of failure are anticipated, understood, and guarded against in balanced and cost-effective ways. Unsafe operating conditions are prevented by design. The product is sized to operate at full capacity for appropriate lengths of time without danger or uncertainty. Humans and the environment are protected from harm, danger, and other hazards. Ideally, the design inherently avoids the possibility of danger or harm. When this is not possible, special protective systems or warnings are added to ensure safety.

Design Axioms

A study of many successful and some not so successful designs made by Nam Suh and his colleagues at MIT led them to propose a set of hypothetical maxims of good design (Suh, Bell, and Gossard, 1977). Analysis and refinement showed that good design comes down to two fundamental propositions. The first is that each functional requirement of a design should be satisfied independently by some aspect, feature, or component within the design. The second is that good designs maximize simplicity; that is, they provide the required functions with minimal complexity. These two propositions are stated in declarative form as the following design axioms (Yasuhara and Suh, 1980), although they might be better termed "commands" for that is what the declarative form implies.

Axiom 1: In good design, the independence of functional requirements is maintained.

Axiom 2: Among the designs that satisfy axiom 1, the best design is the one that has the minimum information content.

An axiom is a statement that is accepted as being true because no counter example has ever been found. The design axioms apply to all design decisions, whether they are for products, processes, systems, software, or organizations. To use the design axioms, first identify the functional requirements (FR's) and constraints. Each FR should be specified such that the FR's are independent of each other and are neither redundant nor inconsistent. It is also useful to order the FR's in a hierarchical structure, starting with the primary FR and proceeding to the FR of least importance. Once the functional requirements and constraints are specified, proceed with the design, applying the axioms to each individual design decision.

Beginning with conceptual design and following through each stage of design realization, the design axioms can be used to provide insight into the design problems at hand, show the way to good design, and form the basis for quality design decisions. Each decision should be guided by the axioms and must not violate them. A product designed by following the axioms will be one that is optimized with respect to all aspects of the lifecycle as well as one that facilitates global optimization of the manufacturing system. Everything that is discussed in this book can be derived and/or explained using the design axioms. They underlie all aspects of DFM.

Illustrative Example 2.1

To illustrate the use of the Axioms, consider the alternative radar antenna mount designs shown in Fig. 2.1. The problem is to design an antenna mount that supports the antenna and prevent movement of the antenna due to external disturbances. Which design is preferred? From the problem statement, we see that the design has two functional requirements:

FR1 Support the antenna.
FR2 Prevent movement due to external disturbances such as wind.

Design A satisfies both requirements by using a rigid tower structure to support the antenna and protect it from unacceptable movement due to wind induced forces. The design lacks clarity, however, because both functional requirements are satisfied by the tower, making tower behavior with respect to each functional requirement ambiguous and hard to predict. In the parlance of the independence axiom, the functional requirements are

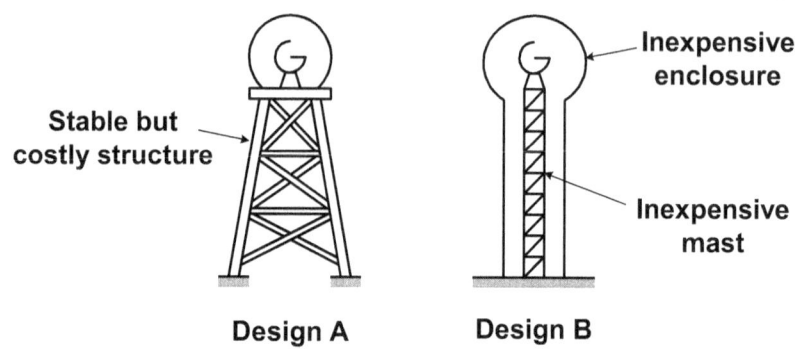

Figure 2.1 Alternative radar antenna mount designs.

"coupled" and therefore dependent on each other. As a consequence, the tower must be carefully designed and constructed using engineering analysis and skilled construction workers. In addition, the tower structure is not robust against unanticipated surprises. For example, if the tower were located near a railroad, its ability to prevent movement due to a passing freight train would be compromised. This is because any attempt to isolate the tower from ground induced disturbances would affect the tower's behavior in the wind. This undesirable interaction defies an easy fix.

Design B accords with the design axioms. Each functional requirement is satisfied independently by using a simple mast structure to support and a barn-like structure to protect the antenna from the wind. This design provides clarity because behavior with respect to each functional requirement is inherently predictable and unambiguous. The mast is easily designed or better yet, purchased from a supplier, because it only needs to provide support. The enclosure can be built using readily available and inexpensive materials by local non-skilled labor. If needed, a simple vibration isolation mount can be used to isolate the mast from ground induced vibration should the need arise. Because each functional requirement is satisfied independently, the use of an isolation mount does not "ripple" to the barn structure and affect its functionality.

Both design alternatives provide the support and protection needed, but Design B is preferred because it has inherent clarity, is robust, and has low information content making it easy to design and construct. Importantly, it will function reliably and safely because failure of either the mast or the barn will not affect the other.

As this example illustrates, maintaining functional independence and minimizing information content yields many benefits:

Good Design

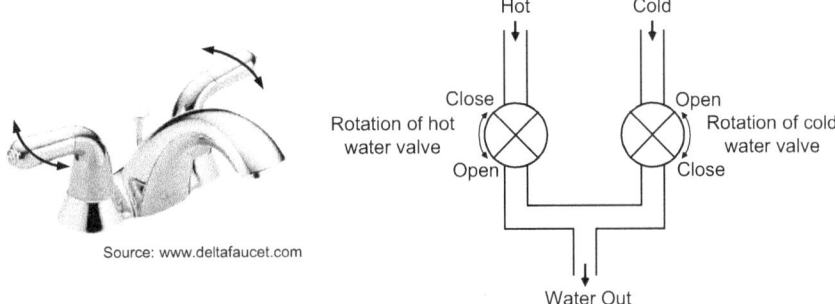

Figure 2.2 Traditional water faucet design.

- Each functional requirement is predictable and unambiguous.
- The design is robust against hard-to-control variation, unanticipated surprises, and other disruptive factors and events.
- The "ripple effect" that often occurs due to "engineering change" is short circuited. This helps prevent sub-optimal design that would otherwise result from engineering change made late in the project.
- The information content of the design can be globally rather than locally minimized.
- Safety and reliability are enhanced because catastrophic failure and/or gradual degradation in capability of one functional requirement does not ripple to or affect capability of the other functional requirements.

Illustrative Example 2.2

To illustrate the axiomatic design approach, consider the design of a simple water faucet for controlling the flow of hot and cold water (this example is adapted from Suh, 2001). A traditional water faucet uses two valves, one for hot water and the other for cold (see Fig. 2.2). Although this design works well and has been used for many years, it lacks clarity because both valves must be adjusted in a trial and error manner to control both flow rate and temperature. In terms of Axiom 1 (the independence axiom), control of water flow and water temperature are "coupled" resulting in an undesirable interaction. Use the design axioms to decouple and simplify this tried and true design?

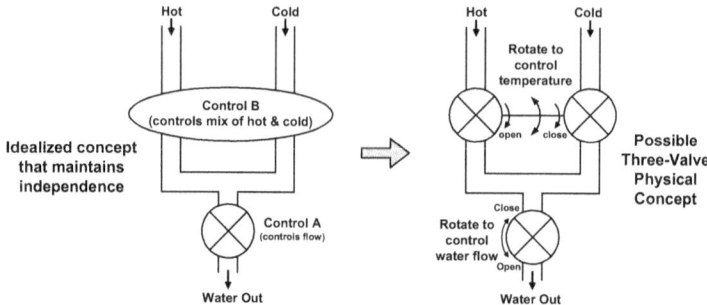

Figure 2.2 Proposed water faucet redesign.

The axiomatic approach is applied in the following three steps:

1. Determine the critical functional requirements. These are the FR's that are either interacting undesirably or are key for implementing the design's functionality.
2. Synthesize a physical concept that maintains independence of FR's.
3. Simplify the physical concept by minimizing information content.

Step 1: The critical FR's are determined by understanding that control of water flow and temperature are coupled in the traditional design.

FR1: Control water flow rate without affecting water temperature.
FR2: Control water temperature without affecting water flow.

Step 2: Synthesize a physical concept that maintains independence of the critical functional requirements. An initial design is proposed as shown in Fig. 2.3 in which "Control A" regulates water flow by opening or closing an outlet valve and water temperature is controlled by "Control B". To increase water temperature, Control B is rotated to simultaneously open a hot-water valve and to close a cold-water valve. To decrease water temperature, Control B is rotated in the opposite direction to simultaneously close the hot-water valve and open the cold-water valve. Since water flow and water temperature are controlled independently by Controls A and B, the independence of functional requirements is maintained.

Step 3: Simplify by applying the information axiom (Axiom 2). Examination of the physical concept proposed in Fig. 2.3 shows it to be unacceptable because it uses three valves while the traditional design only uses two and is therefore simpler and less costly to manufacture. Applying Axiom 2 (the information axiom), the initial design is simplified using the iterative design-analyze-redesign DFM approach (see Chapter 5) by first eliminating one of the valves, then replacing the two valves with two sliding

Good Design 19

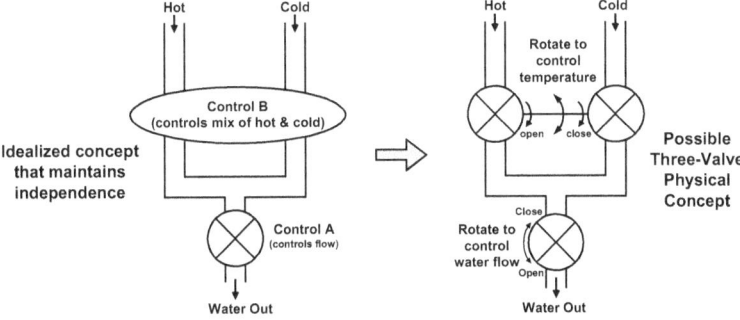

Figure 2.3 Proposed water faucet redesign.

plates, and finally replacing the two sliding plates with one multi-motion (sliding and rotating) plate. The iterative sequence of design simplifications is shown in Fig. 2.4. A commercially available water faucet that uses a ball and socket configuration in place of the sliding and rotating plate is sold by Delta (www.deltafaucet.com) and is shown in Fig. 2.5.

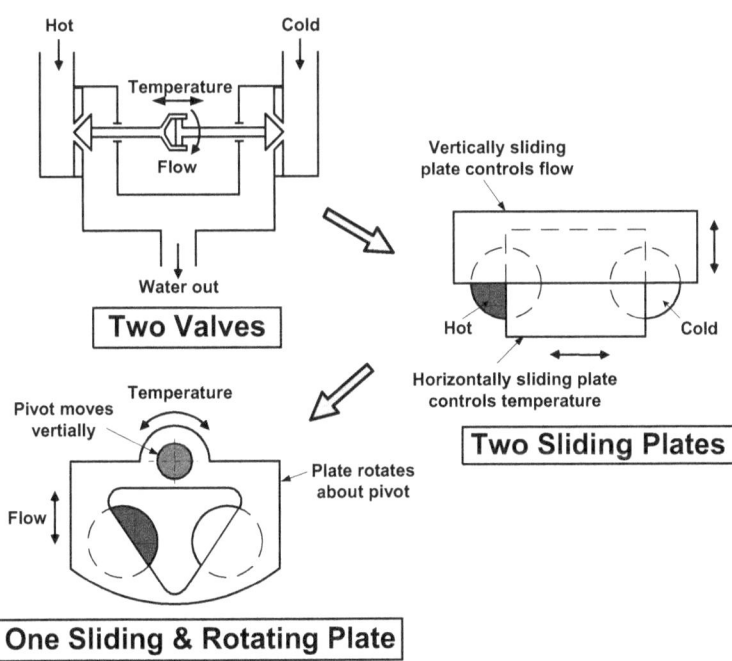

Figure 2.4 Axiom 2 (the information axiom) is used to iteratively simplify the initial design proposed in Fig. 2.3.

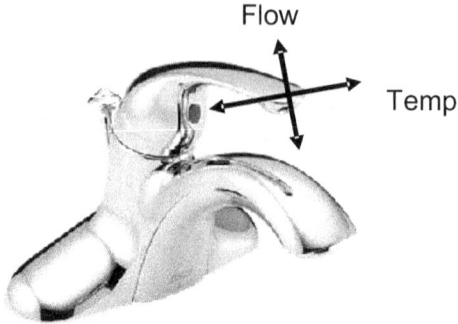

Figure 2.5 Commercially available faucet that satisfies both design axioms.

Application of the design axioms to the analysis and design of products and manufacturing systems is not always easy or straightforward. Because the axioms are quite abstract, their use requires considerable practice. For this reason, the reader is encouraged to study the axioms carefully and to experiment with their use in understanding problems with which he or she is familiar. As confidence grows, the reader is further encouraged to try to apply the axioms to a design or manufacturing problem they are currently working on. By practicing the "axiomatic approach" in this way, one will quickly discover the power of the method and learn how to effectively use it to both guide and evaluate designs.

Metrics of Good Design

The manufacturing success produced by good design is reflected in three high level metrics: total quality, total cost, and total time. Improving these metrics as much as possible is what design for manufacture is all about. Like the concept of good design itself, these metrics underlie all aspects of DFM.

Total Quality

Total Quality is the totality of features and characteristics of a product including the product's design, manufacture, distribution, sale, service, use, and disposal that bear on its ability to satisfy stated or implied needs. It includes a broad range of concerns that pertain to all aspects of the product. These include: (1) how well does the product satisfy the end user needs, (2) how well does the product perform over time, during use, and with respect to safety and environmental impact, and (3) how easy is the product to design, manufacture, distribute, sell, service, and support in the field. In other words, what will cause the customer to select and purchase the

Good Design

product? What will delight, astonish, and satisfy the customer as an owner and user of the product? And, what will make it worthwhile for the manufacturing enterprise to design and sell the product? Total quality, therefore, not only contributes directly to the firm's bottom line by generating demand and a higher selling price, it also makes manufacture and lifecycle support easier and less costly, thus increasing overall profit.

Total Cost

Total Cost is the sum of all costs, both direct and indirect, that result from the design, manufacture, distribution, sale, service, use, and disposal of the product over its lifecycle. All design decisions, both large and small, impact total cost in one way or another. Traditionally, only direct manufacturing costs have been considered when evaluating the goodness or acceptability of a design decision. *Direct cost* is typically computed as the sum of variable and fixed costs. Variable costs depend on production quantity and include material cost, labor cost, and cost of production resource usage while fixed cost typically includes tooling and equipment investments. Use of direct cost in design decisions is convenient because these costs can be estimated and tracked relatively easily. The problem is that this approach ignores the indirect cost that each design decision creates. Often, indirect cost can be many times greater than direct cost.

Indirect cost includes all those costs such as overhead and administrative costs, design cost (time and equipment), and systems "debugging" and problem solving costs, that can't be easily attributed to any one aspect of the design and are therefore lumped together and accepted as the "cost of doing business". Design related indirect cost is composed of three components: systems cost, development cost, and time cost. *Systems cost* includes indirect operations and overhead cost associated with manufacture, distribution, marketing, service, and disposal of the product. Examples include costs associated with purchasing, supplier relations, order processing, inspection, material handling, receiving, shipping, service parts inventory, catalogs, design and manufacturing documentation, worker training, and so forth. *Development cost* includes the direct cost of designing the product and launching it into production plus all of the overhead and equipment cost associated with CAD/CAM/CAE systems, testing labs, prototype shops, R&D labs, and other design support systems that facilitate the design realization process. Finally, *time cost* can be thought of as the cost of lost sales and market share incurred as a result of being late to market or second

Table 2.1 Some Major Components of Total Time

Design Cycle	Order-to-Delivery Cycle	Operational Life
Design & Dev. Time	Order Processing Time	MTBF
CAD/CAM Time	Manufacturing Lead-Time	MTTR
Prototype Fab.	Piece-Part Cycle Time	MTTF
Test/Validate/Fix	Inspect/Test/Calibrate Time	Service Time
Tool Design/Prove	Setup/Tool Change Time	Service Interval
Launch/Ramp-Up	Material Handling Time	
	Assembly Time	
	Delivery/Installation Time	

in the market with a new product. Ideally, it would be nice to be able to click on an icon in a CAD system and receive immediate feedback on the total cost consequence of a design decision. Unfortunately, accurate total cost models and the cost structures required to develop them do not exist in most companies. Chapter 6 presents a powerful alternative for estimating total cost.

Total Time

Total Time is the composite of all times that are affected or determined by design and manufacturing decisions, including design and development time, order-to-delivery time, operation time, useful life, and service and maintenance time. Each design decision, both large and small, contributes in some way to total time. This is illustrated by Table 2.1, which lists some of the many components of total time. Because manufacturing is a repetitive process, cycles occur at all levels and phases of the production process ranging from the new product introduction cycle to the production cycle of a plastic injection molding process or machining operation. Less obvious cycles include tool changeover and machine setup cycles. No matter the cycle, reducing cycle time is a key DFM goal. Shorter cycles mean less quality risk, less cost, less waste, less non-productive activity, and happier customers.

Total Design Value

When used as an adjective, the term "good" means better or best. In the context of design, "better or best" implies a "degree of excellence" or "value" that can be assigned to the design. In other words, a "good design" is one that is considered to have high "value". In design and manufacturing, value is defined as the ratio of function to cost. If the metrics of good design are used to represent function and cost in the value equation, the "total design value" of a design can be expressed as,

$$Total\ Design\ Value = \frac{Total\ Qualty}{Total\ Cost \times Total\ Time} \qquad (2.1)$$

Good design is achieved when total design value is fully maximized, that is,

$$Good\ Design = Total\ Design\ Value \rightarrow Maximum \qquad (2.2)$$

Total design value is qualitative not quantitative. Each term in Eq. (2.1) is likely to be coupled with the others (e.g., ease of serviceability increases total quality while also decreasing total cost and total time). Total cost and total time are multiplicative in recognition of the self-compounding and synergistic effects of DFM noted in Chapter 1. These ambiguities and couplings are not important, however, if the goal is to achieve good design by maximizing total design value as implied by Eq. (2.2). Anything that increases total quality and/or decreases total cost and/or decreases total time is helping to maximize total design value thereby boosting the "goodness" of the design as a result. Therefore, to design for manufacture means to design for maximum total design value. In other words, design for manufacture and maximum total design value are two sides of the same coin.

Summary of Key Concepts

- ➢ Good design is an ideal toward which to strive.
- ➢ A good design fulfills its purpose in the most effective and efficient way possible. It embodies inherent clarity, simplicity, and safety.
- ➢ A good design maximizes total design value.
- ➢ Use of the design axioms, by definition, leads to good design, which in turn results in a design that has been optimized for ease of manufacture, assembly, and lifecycle support.
- ➢ The design axioms underlie all aspects of DFM.

Chapter 3
The Design-Manufacture Interface

The design-manufacturing process can be defined as the transformation of raw materials into finished products that satisfy well-defined customer needs. Next to the end-user, no customer is more important than manufacturing because of its impact on total quality, total cost, and total time. Most design related manufacturing needs arise at the design-manufacture interface because this is where the design and manufacturing process come into direct contact and interact. In this chapter, the many aspects of the design-manufacturing interface are examined with special emphasis on the way DFM can help smooth and simplify the interplay between design and manufacture. Knowledge and understanding of the complex relationships that exist at the design-manufacture interface are essential for effective DFM at all levels of design.

When applied to the manufacturing floor, the basic metrics of good design (see Chapter 2) are more conventionally defined as:

- *Yield*: the ratio of acceptable product to total product produced. An acceptable product is one that satisfies all required tolerance and functional specifications. The goal is to <u>maximize</u> yield.

- *Manufacturing Cost*: the direct cost of performing a processing operation, or sequence of processing operations, in a repetitive process. Always, the goal is to <u>minimize</u> manufacturing cost.

- *Cycle Time*: The average time that a worker or machine or production process spends performing productive work between completion of successive operations, or a sequence of operations, in a repetitive process. The goal is always to <u>minimize</u> cycle time.

These metrics underly almost all design and manufacturing decisions. They guide the design-manufacturing process and define continuous improvement objectives and measures.

Basic Manufacturing Processes

Most design-manufacture interactions involve manufacturing processes that convert starting materials into finished products. From a DFM perspective, manufacturing processes are separated into two types: (1) piece-part processes and (2) assembly processes. Piece-part processes involve the production of individual parts that are homogenous (have the same material properties everywhere) and indivisible. Assemblies, on the other hand, are combinations of several designed parts and purchased components held together by various joining methods.

Piece-Part Processes

Piece-part manufacturing processes generally fall into four distinct categories: (1) shape replication, (2) shape generation, (3) non-shape changing processes, and (4) additive manufacturing processes. Within each of these categories, the piece-part manufacturing process that is selected is largely dependent on material, part geometry, and production quantity.

Shape Replication: In *shape replication processes*, the mass of the starting material is essentially equal to the mass of the final part. During the process, the starting material is "forced" to assume the surface shape of the tool cavity thereby replicating the shape information stored in the die or mold surface. *Near net shape* processes are shape replication processes that produce parts which require little or no secondary processing (e.g., machining) to complete the part. Shape replication processes in general, and near net shape processes in particular, are highly desirable because little or no material is wasted, and the final part can be produced quickly and consistently. To be economically feasible, however, large production quantities are typically required to offset tooling cost. Example shape replication processes include plastic injection molding, casting, powder metallurgy, forging, sheet metal forming, to name a few. The cost per part of most shape replication processes can be calculated as follows:

$$Cost/Part = \frac{C_T}{N} + VC_M + \frac{C_H t_{cycle}}{Y} \tag{3.1}$$

where C_T = total tooling cost (\$), C_M = material cost (\$/in3), C_H = machine cost (\$/hr), N = lifetime number of parts made using the tool, V = part volume (in^3), t_{cycle} = time to process one part (hr), and Y = process yield (usable parts/N). Therefore, the DFM focus is on minimizing tooling cost, material cost, and cycle time, and on maximizing process yield.

Shape Generation: A *shape generation process* is one in which the shape of the part is determined by the pattern of relative motion between the tool and workpiece. For these processes, the mass of the starting material is always larger than the mass of the final part since forming takes place by removal of material. The complex patterns of part and/or tool movement typically result in relatively long cycle times and material waste. In addition, machine setup can be time consuming and wasteful. For these reasons, shape generating processes are, in general, less desirable for high production. At the same time, shape generation processes such as machining are capable of high precision and are often used in secondary processing operations to finish machine castings, weldments, and other less accurate high production parts. When tight dimensional tolerance is needed, they are often the only option. Modern computer technology has helped make shape generation processes like machining highly versatile and flexible. For example, the use of numerical control (NC) makes it possible to change over from one part to another simply by downloading a different tool path program. CAD/CAM systems allow the rapid generation of the NC program directly from the solid model database. And, by using modern sensor technology and feedback control techniques, NC control systems can be designed to give extremely high accuracy. DFM for these processes generally focuses on reducing setup and processing time and on controlling process variation.

Non-Shape Changing Processes: These processes generally involve either property-enhancing processes or surface processing operations. In general, these processes do not alter the shape of the part, except unintentionally in some cases. The most important property-enhancing processes involve heat treatments, which include various annealing and strengthening processes for metals and glasses. Surface processing operations include (1) cleaning, (2) surface treatments, and (3) coating and thin-film deposition processes. Degreasing, sand blasting, shot peening, electroplating, anodizing of aluminum, and organic coating (better known as painting) are common examples. There are a number of good reasons for using these processes including (1) preparation for subsequent processing, (2) wear resistance and surface strengthening, (3) corrosion protection, and (4) color and appearance. From a DFM standpoint, however, these processes should be avoided whenever possible because they are expensive, time-consuming, and typically require special facilities and extra material handling and storage. Consider painting for example. In most cases, painting must be done in a special area that facilitates cleanliness and ventilation. This almost always means that painting cannot be performed on the production line. The result is extra material handling, non-productive waiting and storage, and heightened quality risk.

Additive Manufacturing: In *additive manufacturing processes*, mass of the "build" or starting material is increased as the result of the manufacturing process. Typically, the mass of the final part is approximately equal to the sum of the masses of the components that are added. Parts made using additive manufacturing processes are solid, have essentially the same material properties everywhere and cannot be disassembled. Examples include built-up parts such as weldments and riveted structures. *Layered manufacturing* is a relatively new type of additive manufacturing process in which components are manufactured by building up very thin layers of material, usually in a vertical stack, which are glued, fused, welded, melted, or otherwise joined together to create the solid part. Most layered manufacturing processes, also known as 3-D printing, rely on computer numerical control to determine the shape, configuration, and internal features (e.g., holes and other cavities) of each layer. Layered manufacturing has drastically changed the way prototype parts and products are made and is revolutionizing the ability of inventors to make parts and test their ideas. Also, in some industries, layered manufacturing processes are rapidly becoming a highly practical means for low quantity production of complex parts. For example, some modern rocket engines now utilize 3-D printed nozzles and other complex, hard to make parts. In addition, layered manufacturing has dramatically reduced the time and cost of fabricating tooling such as sand-casting patterns, thereby making these processes more viable for low production. As with all other manufacturing processes, design affects the quality and functionality of the 3-D printed part. Selection of the layering direction, for example, can affect part strength and build time.

Assembly

Assembly is a manufacturing process in which parts are oriented, inserted, and secured to the build in successive steps. Parts are secured using either non-permanent or permanent methods. Non-permanent include clamping, screwing, joining with nuts and bolts, and other mechanical fastening methods that allow disassembly without the need to damage the assembly or any of its component parts. Permanent joining methods, on the other hand, cannot be undone, so the use of permanent joining methods in assembly essentially transforms the build into a unitary part. In this respect, there is considerable overlap between additive manufacturing and assembly. Permanent joining processes include mechanical joining such as rivets and shrink fits, brazing, soldering, and adhesive bonding, as well as all the different welding processes such as resistance welding (spot, seam, projection, etc.), electric arc welding (MIG, TIG, etc.), gas, and solid state welding (friction, ultrasonic, pressure, etc.).

Assembly imposes constraints on the design. Not only must the parts be designed so that they can be assembled and joined together to provide the needed function, they must also be designed so that they are easy to handle, insert, retain, and verify that they have been assembled correctly. Because assembly is an integrative process, problems with detail part designs often surface when they are assembled. Parts don't fit together properly, tools can't reach in the space provided, parts can be incorrectly assembled, and so forth. These problems can often require extensive redesign resulting in costly schedule slippage and undesirable design compromises.

Recognition of the importance of assembly as a design consideration has resulted in a greatly increased emphasis on assembly in the design process. Design and manufacturing practice now focus on eliminating variability and randomness from the process, and on making non-productive operations such as orienting and handling as simple and easy to perform as possible. Design for assembly, which is discussed in Chapters 9 and 10, seeks to eliminate parts and to design the parts that remain to be easy to assemble.

Design and Manufacturing Interactions

The design-manufacturing interface generally involves two different ways in which design and manufacturing interact: (1) manufacturing needs and requirements dictate geometric details of the design, and (2) design decisions directly impact the ease of manufacturing. As shown in Figure 3.1, the interface involves eight major design-manufacturing interactions. These are:

1. Production quantity
2. Product variety.
3. Manufacturing processes.
4. Material and process selection.
5. Tooling.
6. Material handling.
7. Production lines.
8. Mechanization and automation
9. Flexible manufacturing systems.

Production Quantity and Product Variety

Production quantity and product variety both have a profound effect on design and manufacturing decisions. This is because they establish the boundary conditions and initial conditions under which the production system must operate and evolve. How these over-arching decision drivers

The Design-Manufacture Interface

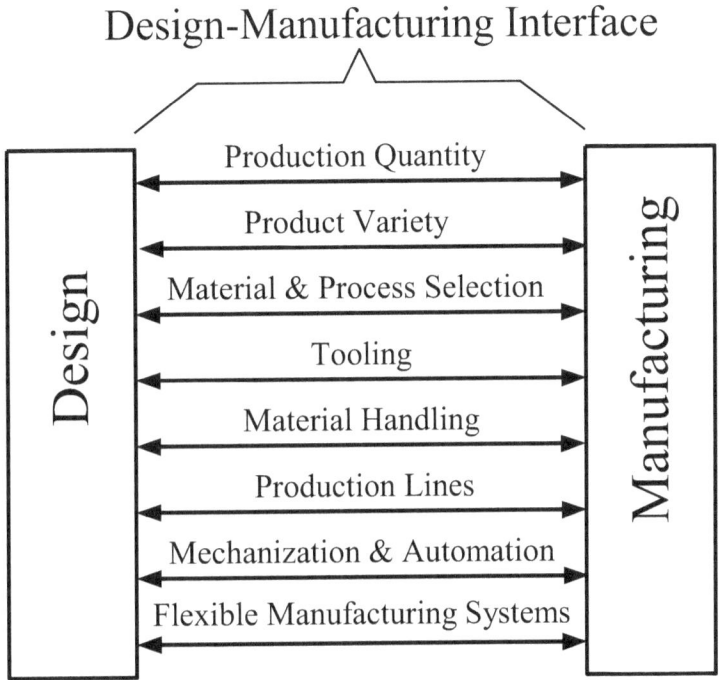

Figure 3.1 Major design-manufacturing interactions.

are accommodated by design and manufacturing choices often determine bottom line profitability and, in many cases, may significantly affect the long-term viability of the enterprise. The ability of the manufacturing system to change and adapt to changes in these parameters is of special importance. Often, DFM is the key to making the manufacturing system robust against such change (see Chapter 13) or accommodating of such change (see Chapter 14).

Production Quantity

Production quantity (also sometime referred to as production volume) is the number of units produced per time-period (typically per year). For most situations, production quantity dictates what materials, manufacturing processes, and production facilities can be considered. For example, if the production quantity for a metal component is large enough, near net shape processes such as die casting may be economically feasible. If this is the case, then the material selected must not only satisfy functional requirements, it must also be suitable for die casting. If, on the other hand, the quantity is small, then a more cost-effective alternative may be to machine the part from metal bar stock. In either case, the decision will depend on existing manufacturing capability, in-house expertise, and resources required and will no doubt produce lasting effects on the business.

Different companies and industries interpret production quantity differently. Typically, however, an annual production quantity in the range of 1 to 100 units is considered to be "low" production, 100 to 10,000 units per year is "medium", and greater than 10,000 units per year is generally accepted as being "high" production (Groover, 2001). Production quantity is an extremely important consideration in determining the mix of manual, mechanized, and automated methods and processes that are used.

Product Variety

Products differ in many ways. They have different shapes and sizes. They perform different functions. From a manufacturing standpoint, these differences constitute what is commonly known as *product variety*. Product variety can be classified as "hard product variety" or "soft product variety" (Groover, 2001). *Hard product variety* is characterized by incompatible differences such as the need for different production processes or tooling, or assemblies that share few common parts, or products that serve different purposes or markets. Conversely, *soft product variety* is characterized by differences that are small and compatible. Consider an ordinary nail, for instance. Nails come in different diameters, lengths, and head styles. Because each nail diameter requires a different diameter feed stock and die diameter, nail diameter is an example of hard product variety. At the same time, once setup for a given nail diameter, the nail making machinery can usually be adjusted to produce nails having different lengths and head styles. Therefore, nail length and head style are examples of soft product variety. Similarly, a four-door sedan, sport utility vehicle (SUV), and pickup truck would be classified as hard product variety, while different SUV models having different trim levels and options would be classified as soft product variety because each SUV model can be assembled on the same line.

With this understanding of product variety, it is evident that, as much as possible, soft variety is to be preferred since this almost always allows more to be done with less. This explains the tendency toward flexible manufacturing approaches and the desire to transfer as much product variety as possible to software control. In the DFM approach, the goal is to find the best balance between hardware, software, and perhaps "human ware". This is best accomplished when the design and production system design are considered concurrently.

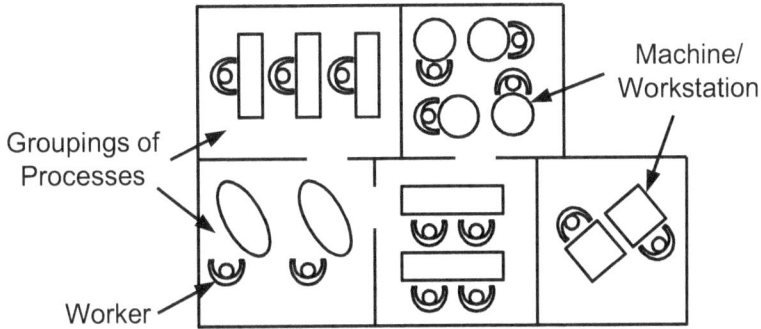

Figure 3.2 A process layout allows process planning flexibility at the expense of ease of material handling and storage. Traditionally, it has been favored by firms that utilize batch production as their primary type of production.

Types of Production

The type of production utilized depends on production quantity and product variety. In general, types of production can be classified as: job shop, batch, cellular, and transfer line. In *job shop production*, low quantities of specialized and customized products are produced. Customer orders are often special and repeat orders may never occur. Equipment is general purpose and the labor force is highly skilled. In *batch production*, one product is made, after which the facility is changed over to produce a batch of the next product. Batch production generally involves low to medium production quantities and is commonly used in make-to-stock situations where inventory that has been gradually depleted by demand is replenished. Equipment is typically arranged in a "process layout" in which production machinery is grouped according to functionality (see Fig. 3.2). The process layout has fallen out of favor in recent years because of the setup required for product changeover, and the excessive material handling it necessitates.

Cellular manufacturing is the preferred alternative to the "process layout" for medium range production when product variety is soft enough to eliminate extensive changeover between one product type and the next. Cells typically consist of several workstations or machines grouped together so that similar parts can be made on the same equipment. In addition to avoiding lost time to changeover, cells are usually laid out for minimal material handling. Fig. 3.3 illustrates a typical *cellular layout*.

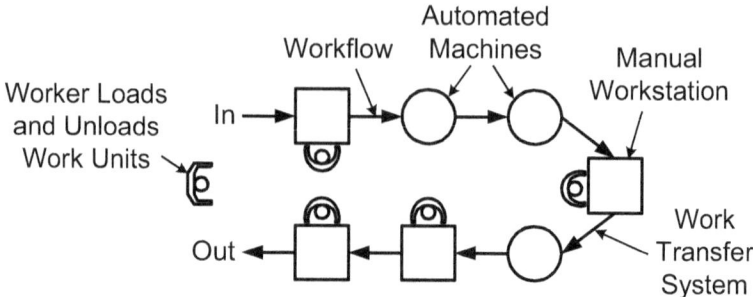

Figure 3.3 Representative cellular factory floor layout. Note that the cell utilizes both automated and manual workstations together with manual and mechanized material handling.

Transfer-line production is typically used in manufacturing situations that involve medium to high production quantities. In this form of mass production, multiple workstations are arranged in sequence, with the partially completed work unit moved from station-to-station to complete the production process in incremental steps. Examples of alternative production line layouts are shown in Fig. 3.4. Each workstation on the line is typically optimized to maximize efficiency and may involve a mix of manual processes, flexible automation, and hard automation. The work is often moved between stations by a powered conveyor or other mechanized material handling method.

Two situations typically arise depending on product variety. When there is no variation in product, a *single model* dedicated production line may be used. This type of production line might consist of just one dedicated machine such as a punch press, or it may consist of several workstations connected by a mechanized material transport system such as a rotary indexing machine (Fig. 3.4). When high mass production is required, two or more single model lines may be operated in parallel (Fig. 3.4).

The second production situation occurs when soft variety is involved. In these cases, the line may be set up to produce several different product models interchangeably and in quantities of one or more. These types of production lines are frequently referred to as *mixed-model* lines. Modern automobile assembly where different models and option packages are assembled on the same production line is an example.

The Design-Manufacture Interface

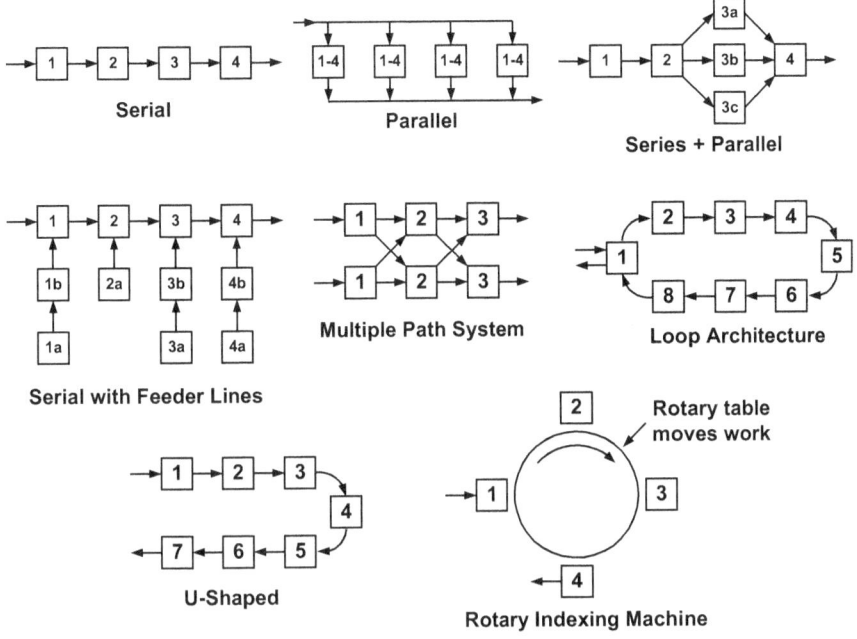

Figure 3.4 Alternative production line architectures.

Material and Process Selection

Depending on the design situation, material and process selection can be a key consideration in achieving the goals of the design. For example, part count, assembly complexity, and secondary processing cost can be greatly reduced by integrating parts together using near net shape manufacturing processes such as plastic injection molding, casting, or powder metallurgy. At the same time, the selection process is complicated by the interactions between material, process, and part geometry that is involved. For example, a powder metal part cannot be designed with an undercut that prevents proper powder compaction or removal from the die. Ultimately, material and process selection come down to trade-offs made between manufacturing cost, investment cost, performance and quality requirements, and tooling lead time.

The process used to select materials and manufacturing processes varies depending on the design situation and on the constraints involved. In general, the process involves progressively narrowing from a large universe of possibilities to one specific choice. In some cases, the possible choices may be predetermined or implied. For example, the bodies of most passenger vehicles are built-up from spot-welded sheet metal. In some cases, however,

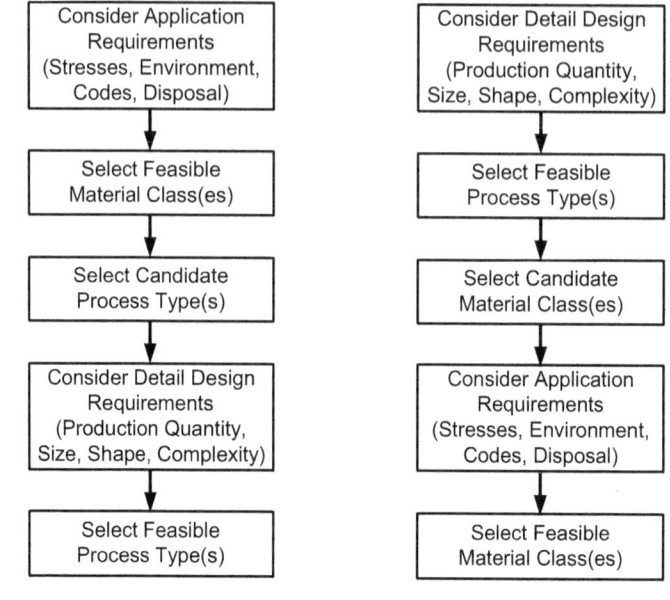

Figure 3.5. Alternative approaches for selecting material and process classes (Dixon and Poli, 1995).

alternatives such as molded plastic or composite materials may be considered. In the ideal, the choice of material and process is wide open, making production quantity, functionality, and in-house manufacturing capability and expertise the main determinants of the final choice.

Two alternative approaches for narrowing the field of possibilities are generally available. In the "material-first" approach (Fig. 3.5a), the material class (e.g., metal, plastic, ceramic) is selected first by considering the material property requirements of the application. Manufacturing processes that are compatible with the selected material class are then evaluated and selected based on considerations of production quantity and on component size, shape, and complexity. In the "process-first" approach (Fig. 3.5b), the process class (e.g., machining, metal forming, casting) is selected first by considering functionality, production quantity, size, shape, and complexity of the component. Materials that are compatible with the process class are then evaluated and narrowed based on the material properties required and the target cost. In both cases, final detail specifications are developed by simultaneously considering both the required material properties and the process requirements and constraints.

Because the universe of possible material choices is large, companies that manufacture material critical products such as medical or food making products encourage the use of formal material selection processes. A formal material selection process involves three main elements: (1) a detailed and comprehensive requirements definition, (2) two or more candidate materials from which to choose, and (3) a formal written set of recommendations supported by a clearly defined selection rational. One way to identify candidate materials and develop selection recommendations is to use a "filtering" process in which the universe of possible choices is first narrowed by considering "must haves" and "key factors" and then developing selection recommendations based on producibility issues. *Must haves* are primary constraints, that is, they are properties or characteristics that the material must possess to be acceptable. *Key factors*, on the other hand, are material considerations that make a given material more-or-less desirable. For example, a "must have" may be a certain minimum tensile strength while a "key factor" may be weight with lighter weight being preferable.

Design-manufacturing issues include a range of practical considerations such as material availability, cost, compatibility with manufacturing processes, ease of joining, aesthetics, and so forth. Because the component functionality, material, and manufacturing process are so tightly coupled, these choices are primary DFM concerns. Getting the design, material and process right is key to a producible design (see Chapter 11).

Tool Design

In manufacturing, a tool is any device or apparatus that enables or facilitates successful conversion of starting materials into finished products. Tool design goals include low total cost, consistent high yield, high production rates, high repeatability from part to part, inherent safety, reliable long-life, and ease of use. Although most product designers may not realize it, when they design a part, especially one that is to be made using a near net shape process, they are actually designing the tool that will make the part since it is the tools shape that is impressed on the workpiece to form the part.

Jigs and Fixtures

Jigs and *fixtures* are tools that hold, locate, and support the work during the processing cycle. These terms are used rather loosely in most industries, but in general, the purpose of a jig or fixture is to ensure interchangeability of the work within the tolerances set by the design. The device may be adjustable or non-adjustable, stationary or portable, and it may have

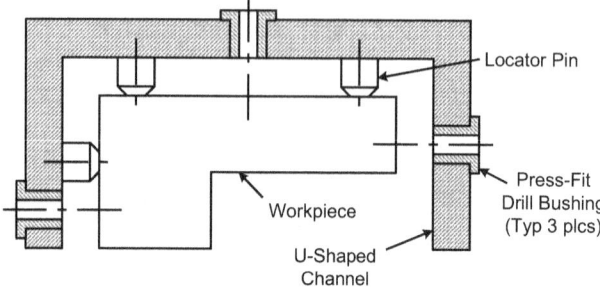

Figure 3.6 Cross-section of a channel jig (clamping details not shown).

numerous physical characteristics depending on the application involved. Most importantly, jigs and fixtures must be designed to facilitate the ease with which the workpiece or assembly is loaded, clamped, and removed.

A *jig* is a device that supports and holds the workpiece at the proper location using "locators". In addition to holding the work in a fixed location, jigs are designed to guide tools in operations such as drilling and reaming to ensure that the operation is done at the exact place and orientation required. For example, the *channel jig* illustrated in Fig. 3.6 was designed to facilitate the drilling of holes into three orthogonal surfaces. Fixtures, on the other hand, are devices that are primarily used to locate the work quickly and accurately, support it properly, and hold it securely. Fixtures vary in design from relatively simple tools to expensive, complicated devices. In addition to machining operations, jigs and fixtures also serve to locate, position, and align parts for weldments and assembly processes. To illustrate, consider the manufacture of a pre-hung door and frame assembly. In this case, a jig is used to locate and hold all components in the correct position and orientation with respect to each other. When all components are properly located and aligned, the parts are welded or screwed together to lock everything in place and in perfect alignment.

Jigs and fixtures must be carefully designed to ensure that the work is properly located and supported. In general, the work to be held and located has six degrees of freedom, three translations, and three rotations. The translations are in the x, y, and z directions of three-dimensional space, and the three rotations, roll, pitch, and yaw, are about the x, y, and z directions, respectively. Ideally, it is desirable to locate and hold the work so that it is exactly constrained by applying just enough constraints to unambiguously define its position with respect to the six degrees of freedom. One way to do this is to design the fixture using the 3-2-1 fixturing principle (see Fig. 3.7).

The Design-Manufacture Interface 37

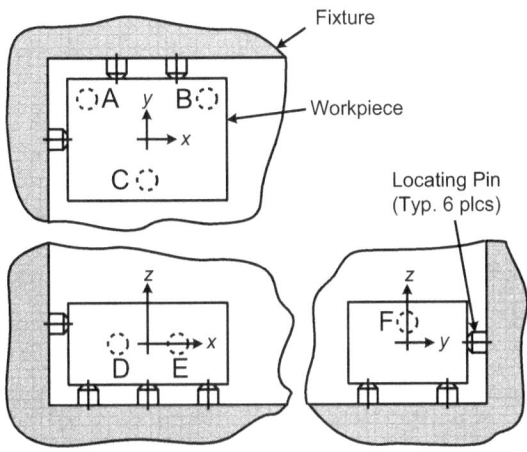

Figure 3.7 Workpiece located using the 3-2-1 fixturing principle.

When the work is set on the *x-y* plane, it contacts at three points (A, B, and C in Fig. 3.7), which removes three degrees of freedom, one translation and two rotations. A second translational freedom and the third rotational freedom are removed by sliding the work along the *x-y* plane until it contacts the x-z plane at two points (D and E in Fig. 3.7). The final translational freedom is then removed by sliding the work along the *x-y* and *x-z* planes until it contacts the *y-z* plane at one point (F in Fig. 3.7).

Once located, the work is clamped in place, usually by elements that are free to move relative to the fixture. Clamping force is applied to the work in a variety of ways including mechanical linkages, cams, eccentric pivots, power screws, and so forth. The clamping means must be carefully designed in conjunction with the locating features to properly react the forces, torques, and pressures that may act on the workpiece during processing. Care must also be taken to ensure that the workpiece is not damaged in any way, either by the clamping mechanism or by the processing forces and pressures. Clamping force can be high and is typically produced by mechanical, pneumatic, or hydraulic means.

Jigs and fixtures are indispensable aids in many manufacturing and assembly situations. The opportunity that DFM offers is to design the product and the jig or fixture as a coordinated system. This allows the jig or fixture design to be simplified by providing needed fixturing features on the workpiece itself. It may even be possible, in some cases, to eliminate the need for the jig or fixture altogether by DFM.

Figure 3.8 Dies and molds transfer shape information to the workpiece (Source: EWT/3DCNC, Inc. advertisement).

Dies, Molds, and Cutting Tools

A cookie cutter and an ice cube tray are simple examples of a die and mold, respectively. Dies and molds work by transferring their shape to the workpiece (see Fig. 3.8). Dies deform and/or shear a solid workpiece by applying pressure and force on the workpiece itself. They are typically used in shape replication processes such as sheet metal cutting, stamping, bending, forming, and drawing as well as bulk deformation processes such as rolling, extrusion, and forging. Molds, on the other hand, create the part shape by constraining flow of a liquified starting material and forcing it to assume the shape of the space formed by the mold cavity and core. This generally requires a material that flows viscously when heated and that solidifies when cured or cooled to room temperature. Molds are used in shape replication processes such as casting and plastic injection molding. Dies and molds are usually made of hardened tool steel or other strong and wear resistant materials that allow them to hold up under extremes of pressure and heat. Die design can be a lengthy, trial and error process because of hard-to-control variation in elastic spring-back, Poisson's ratio, and friction between the die and workpiece surfaces. The same is true for molds for reasons of solidification shrinkage, residual stress, porosity, and other hard-to-control factors.

A chisel is a simple example of a cutting tool. Cutting tools remove material from a solid workpiece such as a metal bar by means of shear deformation. Shape generation processes such as turning utilize a cutter having a single point cutting edge. Milling and drilling tools generally utilize multiple cutting edges, while each grain of abrasive material in a grinding wheel is a microscopic single point cutting edge. In machining processes, cutting speed and depth of cut (feed) determine the production rate and tool life so speeds and feeds are carefully chosen and optimized.

Tool Material Selection Considerations

The principal materials used for tools can be divided into three major categories: ferrous metals, nonferrous metals, and nonmetallic materials. Ferrous tool materials have iron as a base metal and include tool steel, alloy steel, carbon steel, and cast iron. Nonferrous materials have a base metal other than iron and include aluminum, magnesium, zinc, lead, bismuth, copper, and a variety of other metals and their alloys. Nonmetallic materials include woods, plastics, rubbers, epoxy resins, ceramics, and diamonds that do not have a metallic base.

The material selected for a given tool application depends on the mechanical properties required and the desired production life of the tool. Because tools such as molds and dies can be extremely expensive and require lengthy lead-times to fabricate and prove out, trade-offs are often made. For example, by combining CAD technology and 3-D printing with the cast Kirksite tooling process, tooling for prototype, bridge-to-production, and short-run parts can be quickly and inexpensively created. Kirksite is a moderate strength zinc-based alloy that was originally developed for sheet metal forming tools. To create a Kirksite plastic injection mold, the CAD file for the part is used to produce a 3-D printed model in hours with minimal human intervention. The 3-D printed master is then used as a pattern to cast the Kirksite mold cavity to net shape. The resulting mold is capable of reliably producing fully functional plastic parts, but tool life will typically be relatively short. Hence, a costlier Class A tool will be necessary if long-run production of high-quality parts is desired.

Material Handling

Material handling involves the flow of materials to, from, and within the manufacturing workspace. Although material handling adds no value to the product being manufactured, it is an essential and indispensable aspect of manufacturing that is difficult to avoid. It can be capital intensive, especially in high production situations where automation is an option. Even when automated, material handling almost always requires manual labor. Many of the manual tasks to be performed can be mind-numbingly repetitive and in some cases, dangerous and back breaking. For these and many other reasons, material handling represents an important opportunity for cost reduction and cycle time improvement by designing the product in ways that simplify or eliminate material handling.

Material Flows

A *material flow* is the factory floor movement of a specific type of material or object that is required as part of the production process. Product design and production system design decisions often determine what material flows are necessary and how they must be implemented in the plant. By considering material flow early in the design process, some of these flows can be greatly simplified or possibly eliminated altogether. Common material flows are categorized as follows:

1. *Product Flow*: Process path along which the unfinished product travels.
2. *Workspace Flow*: The handling that occurs within the workspace directly prior to processing or assembly.
3. *Supply Flow*: The direct movement from a receiving process to a marshaling area or a workspace with a one-to-one matching between product flow requirements and supply flow requirements.
4. *Hardware Flow*: The direct movement of components from a marshaling area to a workspace. Like supply flow except pieces are not individually accountable because of small size and value.
5. *Trash or Scrap Flow*: The movement of scrap (especially used packaging materials) away from the workspace to a marshaling area or the plant boundary. Rejected and/or defective materials are included in this flow.
6. *Bulk Material Flow*: The movement of loose, unpackaged materials without constant dimensions to the workspace.
7. *Container Flow*: The movement of reusable shipping materials, both containers and dunnage, from the workspace back to a marshaling area or to the plant boundary.
8. *Fixture Flow*: The movement of jigs and fixtures back to the initial workspace for reuse.

Material Handling Equipment

Material handling equipment that comes in direct contact with the product and therefore needs to be coordinated with the product design typically includes production line equipment, and part handling containers.

Production Line equipment: There is a variety of material handling equipment specifically intended for use on production lines and in work

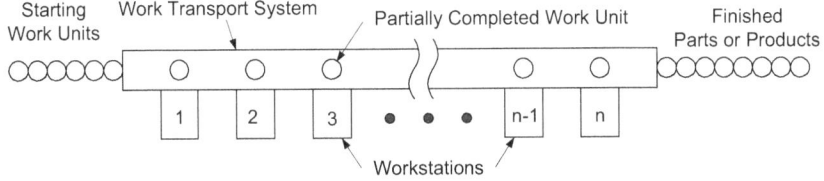

Figure 3.9 A typical multi-station transfer line.

cells. Examples include industrial robots and other parts manipulators, dial indexing tables, transfer mechanisms used in automated flow lines, vibratory bowl feeders, and other parts feeding and delivery devices.

Part Handling Containers: Correct part orientation is often critical for assembly and other manufacturing operations. Achieving the correct orientations is expensive in terms of both cost and time, therefore once correct orientation is achieved, it should be maintained. Material handling schemes that assist and maintain orientation include palletized trays, magazines, tube feeders, part strips, and kits. Often suppliers will package parts in magazines or as part strips. Kits are palletized trays that contain recesses and "nests" that correctly arrange and orient the parts used to complete a given product model.

Production Lines

A production line is any manufacturing system that is setup and operated to repeatedly and reliably transform successive work units from one state to a more advanced state of completion. The defining characteristic of a production line is "reliable repetition". Almost any design can be made in a well-equipped workshop by skilled technicians. The detail features of a design become much more critical, however, when the goal is to mass produce in quantity with the highest possible yield, as fast as possible, and with as little stoppage or interruption as possible. From the perspective of DFM, the production line is where the "rubber hits the road".

A variety of different line layouts, material handling schemes, worker placements, and workstation designs are possible depending on the nature of the product and the manufacturing tasks to be performed. In Fig. 3.9, the production line consists of several "workstations" positioned in a straight-line configuration and connected by a "work transport system". Product manufacture begins with launching of a starting work unit onto the input end of the line. A work carrier or fixture may be used to support and facilitate work unit movement along the line. The work unit travels through each workstation, where workers or mechanized devices perform tasks that progressively complete the work unit until it exits the final station as a finished product.

Work is moved from station to station either manually or using a work transport system. *Work transport systems* are material handling systems or devices that are set up to transfer "in-process" work units from one station to next. There are as many work transport systems as there are manufacturing engineers to invent them. Transfer is usually performed in one of three ways: (1) synchronous transfer, (2) free-transfer (non-synchronous), or (3) continuous transfer. In *synchronous transfer*, the work transport system includes an "indexing" mechanism that causes each in-process work unit to transfer simultaneously from one station to the next. Although more costly, synchronous transfer is typically used when one or more workstations are mechanized. Examples include cellular manufacture and rotary indexing tables. Alternatively, in *free-transfer systems*, the work unit is transferred when work at the station is complete and is not tied in any way to what is happening at the other workstations. Methods of moving the in-process work from station to station can range from using a simple roller conveyor to using a sophisticated robotic transfer system. When the product is relatively large, or heavy, a manual or powered work carrier dolly may be an option. Free-transfer systems usually require some minimal buffer storage at each workstation. In *continuous transfer systems*, the in-process work units are moved along at a constant rate as work is performed. In this case, the workers must perform their assigned tasks as they walk or move along with the moving work unit. Continuous transfer systems are used on high volume manual assembly lines such as automobile and appliance assembly, and on processing lines such as painting or heat-treating lines.

The number of workstations on a manual assembly line does not necessarily equal the number of workers. This is especially true for large products, where it is often possible to assign more than one worker to a station. For example, in final assembly plants that build cars and trucks, two workers, or perhaps two robots, at one station might perform assembly tasks on opposite sides of the vehicle. Multiple manning conserves valuable floor space in the factory by reducing the number of stations.

The purpose of a workstation is to accomplish one or more processing tasks or operations. The number and nature of the operations to be performed at a station are often tentatively decided during overall system design and may be revised often as station designs are optimized and refined. Operations assigned to each workstation depend on numerous factors such as the order in which they must be performed, degree of complexity and skill required, time available, and so forth. Typical operations on a production line include part mating and securing, application of adhesives, use of tools, application of heat, and measuring. Errors may arise from parts fabrication, processing or assembly equipment, jigs, fixtures, part feeders, human performance, and so on. Verification must include determination that the operations and tasks have been performed correctly and that the work has been accomplished within prescribed tolerances on processing forces, accelerations, temperature, pressure, cleanliness, or whatever may be of concern.

Time Considerations

The primary parameter used to design, operate, and measure the performance of a production line is the time required to complete one production cycle. This critical parameter, known as the *cycle time* (T_c), is defined as the average time between completion of successive units in a repetitive process. In general, each workstation on a multi-station line will have its own unique cycle time ($T_{station}$). For a synchronous transfer-line, the cycle time for the line is established by the workstation having the longest cycle time ($max\{T_{station}\}$) since the line cannot cycle until all workstations have completed their work. This critical workstation is often termed the "bottleneck" workstation. The cycle time of a workstation is typically composed of the following three components (Groover, 2001):

- **Operation Time (T_o):** The time that a worker or machine spends performing productive work between completion of successive units in a repetitive process. *Productive work* is any task that adds value to the work unit

- **Repositioning Time (T_r):** The time spent repositioning workers and/or transferring a work unit from one workstation to the next between completion of successive units in a repetitive process. Often an average value for all the workstations on the line is used in design and analysis calculations.

- **Non-Productive Time (T_{np}):** The time spent waiting or performing non-productive tasks between completion of successive units in a repetitive process. *Non-productive work* is work that may be necessary and important for keeping the production line operating but does not directly add value to the work unit.

Summing these components, the cycle time for a typical production line workstation is,

$$T_{station} = T_o + T_r + T_{np} \tag{3.2}$$

Production Time (T_p): The production line cycle time per unit (typically minutes/unit) that is "actually" achieved. When used for the purposes of designing and setting up a repetitive process, T_p is generally understood to be the shortest production line cycle time that is possible in practice. Once the line is up and running, T_p is understood to be the day-to-day cycle time at which the production system is operated. By operating the repetitive process faster or slower, production time can be and typically is adjusted daily by the system supervisor to satisfy demand. Production time is calculated as follows:

$$T_p = T_c/E = (T_r + \max\{T_o\})/E \tag{3.3}$$

where T_c = cycle time, T_r = average repositioning time for a multi-workstation line, T_o = workstation operation time, and E = uptime efficiency (i.e., the proportion of time the line is operating). Uptime efficiency recognizes the fact that, in practice, there will be unavoidable stoppages and interruptions. It is an important guiding parameter used in production system sizing and operation. Typically, it is initially estimated based on past experience with similar lines and is then refined as experience is gained.

Effective Work Time: Most manufacturing companies divide the 24-hour workday into three 8-hour shifts. In most practical manufacturing situations, planned stops are needed and therefore, not all 8-hours in a shift are available for productive work. *Effective work time* is the available work time per shift minus planned stops. *Planned stops* include downtime for preventative maintenance, operator breaks, and other down time such as a morning meeting and end of shift clean up time. Effective work time represents the true amount of time that operators and/or machines have available for touching and making product. There are many factors involved in calculating effective work time, and every organization calculates it differently. No matter how it is done, it is important to be as accurate as possible because effective work time is a factor in many critical production decisions.

Takt Time: This is the time available to complete a work unit to meet a specified annual production quantity. In other words, it is the cycle time in a repetitive process necessary to satisfy a stated demand. All processes that are directly or indirectly related to a production system should follow the same takt time. "Takt" is the German work for "rhythm" or "pulse" and is computed as,

$$Takt\ Time = \frac{Effective\ Work\ Time \times Number\ of\ Shifts}{Annual\ Production\ Quantity/Available\ Workdays} \quad (3.4)$$

There is often confusion between takt time and production time. Although these terms may appear similar, they are actually quite different. *Takt time* is the time required to complete one successive work unit on a repetitive production line to produce a given number of products in a given effective work time. Takt time will not change unless production quantity and/or effective work time change. In practice, unplanned stops and other interruptions occur daily making it usually impossible to operate a production line at a constant takt time day in and day out. For this reason, actual production time may be different each day. In general, most manufacturing companies strive to operate the production line so that, on average, production time equals takt time. If the average daily production time is less than the takt time, over production will occur; if it is greater, production shortages may occur.

Work Content

Work content is the total productive work required to complete the finished work unit. It is expressed as a time, usually in minutes. Work content is determined using different methods. The conventional approach is to perform an industrial engineering time and motion study, ideally under production conditions, in which the time to complete each process task is measured and then summed to determine the total work content. Alternatively, the work content can be subdivided into a series of task elements that have been standardized. Work content is computed by summing the "standard time" for each work element. *Standard time* is the time needed for a well-trained worker to carry out a defined task using an established method and working to a normal speed during a real workday. Operators cannot work at 100% efficiency continuously for their entire shift. Standard time takes this into account along with a variety of other factors including fatigue, operator attitude, rest room breaks, and so forth. Work content is directly tied to the number of parts, the detail part geometry involved in the design, and the number and complexity of the operations to be performed. The reduction in work content time produced by the DFM focus on part count reduction (Chapter 9) and design for assembly (Chapter 10), is one of the most tangible benefits of DFM.

Number of Workstations

When the production volume, effective worktime, number of shifts, and work content time are all known or estimated, the theoretical number of workstations ($n_{theoretical}$) needed on the line to make the desired production rate can be estimated using Eq. (3.4) and the following relationship:

$$n_{theoretical} = \frac{Work\ Content\ Time}{Takt\ Time} \qquad (3.5)$$

To illustrate, suppose a manufacturing company wishes to produce 15,000 product units per year, each having a work content of 20 minutes. Further, suppose that the company works 240 days per year, that it operates one shift per workday, and that effective work time per workday is calculated as,

Available work time = 8.5 hr x 60 min/hr = 510 minutes	
Lunch (not paid)	- 30 minutes
Two coffee breaks	- 30 minutes
Morning meeting	- 15 minutes
End of shift clean-up	- 15 minutes
Effective work time	420 minutes

Takt time is calculated from Eq. (3.4) to be,

$$Takt\ Time = \frac{420\ min \times 1\ shift}{15,000\ units\ /240\ workdays} = 6.72\ min.$$

Substituting the work content time of 20 minutes and takt time of 6.72 minutes into Eq. (3.5) gives a theoretical number of workstations of,

$$n_{theoretical} = \frac{20\ min.}{6.72\ min} = 2.98 \approx 3\ workstations$$

Uncertainties and Inefficiencies

The theoretical number of workstations calculated using Eq. (3.5) is far from representative of actual manufacturing floor reality. This is due, in part, to a variety of uncertainties and inefficiencies.

Production Uncertainties: One hundred percent uptime of a production system is difficult to achieve, even under the best of circumstances. Causes of hard-to-control time loss includes at least the following:

- *Breakdowns*: time that production is stopped for repairs.
- *Setup and changeovers*: time to changeover from one product or product model to another.
- *Idling and minor stoppage*: downtime to correct minor problems.
- *Defects*: time loss caused by part defects and poor quality.
- *Start-up losses*: time lost because the process produces defects until it achieves steady state operation.

Experienced manufacturing engineers will often size manual assembly lines to produce at a cycle time that is roughly 85% of the required takt time. This is called the "85% rule". This rule allows for unpredictable lost time, makes time available for servicing (e.g., tool changes), and permits operators to work at a productive rate that results in properly manufactured and assembled, quality products. The 85% rule is, in essence, an estimate of *uptime efficiency* in Eq. (3.3). This is the percentage of effective work time during which the line is producing acceptable product.

Production Inefficiencies: In addition to line inefficiency, there are numerous unavoidable production inefficiencies. These include the following:

- *Imperfect line balancing*: it is difficult to divide the work equally among workstations.
- *Task time variability*: there is inherent and unavoidable variability in the time required to perform a given task; this is especially true on manual production lines.
- *Repositioning time losses*: the amount of time available for productive work at each station is typically less than the line cycle time because some time will be lost at each station due to repositioning of the work or the worker.
- *Rework*: defective components and other quality problems will cause delays and necessitate rework that adds to the total production line workload. Sometimes called the "hidden factory", rework can significantly increase manufacturing cost while also putting a damper on productivity. The need for frequent rework is a bright red flag signaling that something is wrong with the design.

Line Balancing

Assigning tasks to individual workstations (or workers) so that all workstations and workers have an equal amount of work is a challenge. Typically, the work content can be divided into *minimum rational work elements*, where each element is a small portion of the total work content. In general, the minimum rational work element is the smallest practical amount of work into which the total job can be divided. Different work elements will require different times, and when they are grouped into logical tasks and assigned to workers or workstations, task times will not be equal. Because all time elements are not the same, some workers or workstations will end up with more work, while others will have less. In a synchronous system, the cycle time of the production line is determined by the station with the longest task time. Constraints that limit the ability to group work elements so that the sums (task times) are nearly equal include the following:

- **Production Rate:** the line is designed to achieve a desired production rate. Therefore, the sum of the work element times assigned to each station must be less than or equal to the required production time.

- **Task Sequence:** some elements must be done before others, creating what are called *precedence constraints*. In many situations, balance is impossible without violating a precedence constraint. Minimizing the number of precedence constraints and optimizing the assembly sequence is another way the DFM approach directly benefits the production line.

These and other limitations make it virtually impossible to achieve perfect balancing of the line, which means that some workers or stations will require more time to complete their tasks than others. The inability to achieve perfect balancing results in some idle time at most stations. Because of this idle time (non-productive time), the actual number of workstations required on the line will almost always be greater than the theoretical number calculated using Eq. (3.5).

DFM and Production Lines

Many production line problems can be addressed and solved by the DFM approach in the early stages of design. Dramatic reductions in work content time is a natural outcome. But DFM can do much more. For example, by defining a preferred assembly sequence early in the design, parts can be specifically designed to avoid precedence constraints. In addition, by developing the design and production line concurrently as a coordinated system, rational work elements can be defined early, and the line balanced as part of the design. A key goal of this book is to expand the concept of DFM to one of being a strategic approach for globally optimizing the design-manufacturing process.

Mechanized and Automated Manufacturing

Production processes are often designed and optimized to reduce manufacturing cycle time and, as a result, they may involve manual operations, mechanized operations, automated operations, or a mix of all three. A distinction is generally made between mechanization and automation. *Mechanization* implies that something is done or operated by machinery rather than manually whereas *automation* means a system in which many or all the processes involved in the production, movement, and inspection of parts and material are automatically performed or controlled by self-operating devices. Automation denotes sensing, closed loop control, and some degree of decision making in addition to mechanization. *Flexible automation* includes the added capability of being easily reprogrammed or adapted to meet varying or new production needs.

The degree of mechanization and automation depends to a large extent on the variety of products being produced and on production quantities. If production quantities are low and product variety is extensive, then the production processes are likely to be manual with an emphasis on accommodating variety. A glass blower who makes a wide variety of different ornamental decorations is an example. In this case, the glass blower probably does his work at one work bench with all his tools readily at hand. He or she is ideally set up to make individual pieces, one at a time, and in any order necessary to meet production demand. At the other extreme, if production quantities are large and only one product is made, it is likely that dedicated mechanization or automation would be used. For example, a company that makes millions of a certain type of light bulb each year is likely to manufacture the light bulb using a production process that is fully automated, but capable of only producing that one light bulb.

It is important to note that mechanization and automation impact part and product design in a variety of ways. For example, if a part is to be fed using a vibratory bowl feeder, it must be properly designed for that environment. This implies that the material must hold up under the forces and motions imposed by vibratory bowl feeding conditions. Similarly, if a part is to be handled by a robot, it must have appropriate design features that allow end effector gripping and it must be capable of withstanding the gripping forces involved. Automation and mechanization therefore often superimpose additional constraints on part and product design. In these cases, a "design for automation" approach that considers the needs of the automation early in the design of the product is a must. In addition, it is essential that the systems integrator or supplier of the automation equipment be included as an important member of the design team.

Flexible Manufacturing Systems

Flexible manufacturing systems offer the ability to adapt quickly and easily to changes in product mix and production conditions. Flexible manufacturing is generally manifested on the production floor in the form of programmable automation and new approaches to material handling and part fixturing. A robot represents a typical example of flexible automation. Because it is programmable (under software control), the robot can be programmed to perform a variety of different tasks within its envelope of capability, without change or modification to its physical form. Hence, in a new manufacturing application, instead of designing a special piece of equipment, a standard robot, programmed to perform the new task, can be readily placed on the line. Product design enters the picture in a variety of ways. For example, the design concept must readily accommodate the full spectrum of anticipated product changes. In addition, robotic assembly requires a design that accounts for the special needs of robotic assembly such as top-down assembly and parts that self-guide and self-locate during insertion into the build.

Kitting is an example of a flexible approach to material handling when a variety of product models are involved. A kit is nothing more than a pallet designed to hold parts at known locations and orientations with respect to an established reference frame on the pallet. A different kit, tailored to hold the parts for each product or model variation, is used to supply parts to the assembly worker or programmable automation. In this way, the assembly station is supplied with the right parts, at the right time, and in the right orientation, to build the specific product or model variation required. For manual assembly, the assembly worker needs only to look at the kit to know which product variation is to be built. In the case of programmable automation, the station needs to read a bar code and/or touch a tooling post on the kit pallet to know what product is to be built and therefore what assembly program to run. Again, the design must be coordinated with the kitting concept to ensure success of the kitting approach.

The ability to manufacture to customer order, produce a correct first part, manufacture a variety of different product types or product models in any sequence and quantity, and rapidly introduce a new product or change an existing design are a few of many important capabilities promised by flexible, computer-based manufacturing. But, to achieve these benefits, the product must not only be correctly designed for its intended function, it must also be designed to satisfy the requirements and constraints of the flexible manufacturing system (see Fig. 3.10). The importance of design in making modern flexible manufacturing possible cannot be understated.

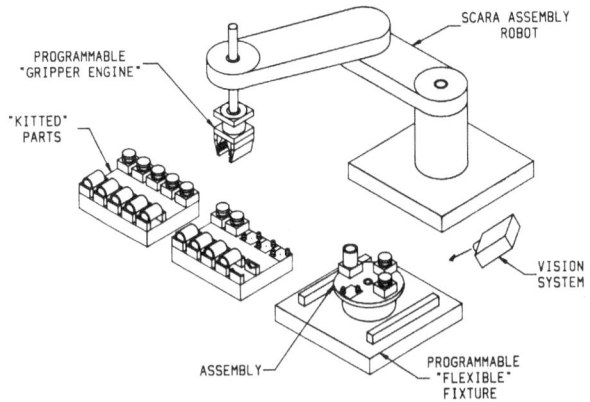

Figure 3.10 The product design must be properly matched to the needs of advanced flexible manufacturing technologies.

Summary of Key Concepts

> The interplay between design and manufacturing is complex and tightly coupled. As the first manufacturing step, design influences or directly dictates many manufacturing decisions. At the same time, manufacturing needs and constraints limit many design choices.

> More than anything else, production quantity and product variety determine the design and manufacturing direction.

> Soft product variety is always preferable to hard variety because more can be done with less.

> Jigs, fixtures, and production tooling are critical DFM opportunities because they directly interface the design with the process.

> Non-productive activities, such as material handling, are only necessary because of the way the product is designed.

> DFM can do much more than just reducing work content such as optimizing the assembly sequence, eliminating precedence constraints, and helping to balance the production line.

> The design must be carefully matched to advanced flexible manufacturing, assembly, quality control, and material handling technologies to fully realize promised productivity gains.

Chapter 4
Design Descriptions

To design a manufacturable product, three different descriptions of the design must be developed. The design is not done until all three design descriptions are complete. Each design description is defined as follows:

1. **The Functional Description:** This is a detailed statement of the design problem to be solved. It defines the intended purpose of the design, details the functions the design is to perform, and states this design specification in precise, measurable terms.

2. **The Form Description:** This description defines how the design is to fulfill these functions in terms of working principle, geometric layout, and individual part configurations including dimensions, tolerances, materials, and manufacturing processes.

3. **Fabrication Description:** Better known as the manufacturing plan, this design description defines how the design and its components are to be made or procured, assembled, tested, and brought to market to be sold.

As discussed in Chapter 1, the "basic actions" of a manufacturing enterprise are divided into four main groupings: (1) decide what customers want, (2) set up the factory to make it, (3) produce it, and (4) sell and support it. The three descriptions of design represent the "work product" of these basic actions. The *functional* description is a statement of "what" customers want. The *form* and *fabrication* descriptions, taken together, tells the manufacturing enterprise "how" to produce, sell, and support the design. Like the "basic activities" of the manufacturing enterprise, each of the design descriptions interact with each other in complex and hard to predict ways. In this chapter, we discuss each design description and then examine how each interacts with the others. Understanding and appropriately managing these design descriptions is key to effective design for manufacture.

The Functional Description

Nam Suh (Suh, 2000) defines design as the "interplay between <u>what</u> we want to achieve and <u>how</u> we want to achieve it." The *functional description* of a design is a clear and exact statement of the "what the design is to achieve" in precise engineering terms. The correct functional description ensures that the right problem is being solved and leads to the one best "winning" design solution. The functional description can be formulated in a variety of ways depending on the design situation and may mean different things to different companies. In general, it includes a mission statement or business plan, some sort of statement of the design problem to be solved, and a detailed design specification that is used to guide the design process.

Design projects typically begin with identification of a market opportunity or un-met user need together with broad constraints and objectives for the project. This information is frequently formalized as a *mission statement* (also sometimes called a *charter* or a *design brief*). The mission statement is typically the result of product planning activities within the firm and it usually specifies the direction to move in but not the destination. Ulrich and Eppinger (Ulrich, 2000) suggest that the mission statement should include a brief (one-sentence) description of the design, key business goals, target market(s) for the design, and assumptions and constraints that guide the design effort.

Building on the mission statement, the *problem statement* is a clear and concise declaration of what the design is to achieve together with carefully chosen focusing assumptions. *Focusing assumptions* limit the scope of the problem. For example, including the term "manually powered" eliminates other means of powering the design from consideration. Such assumptions constrain the design solution and must be carefully chosen to be in harmony with the mission statement and other stated goals of the project. The problem statement establishes the level of the problem and forces the team to carefully consider exactly what the design task involves. It tells the team what to focus on and what aspects of the use and manufacturing environment need to be understood. Most importantly, the problem statement suggests questions that need answering.

The problem statement should not suggest or imply a design solution or otherwise misguide the problem-solving process. How the problem is defined determines the solutions that are proposed. If the problem statement suggests a solution, then the opportunity for finding the best solution is lost. In addition, the problem statement should constantly be questioned to ensure that the right problem is being addressed. Nothing is worse than introducing a new design that "flops" because it doesn't address or resolve the "real" problem.

Any entity that interacts in any way with the design is a *customer*. This includes manufacturing, distribution, use, and disposal as well as testing and code authorities such as Underwriters Lab (UL), the American Society of Mechanical Engineers (ASME), the Environment Protection Agency, and the many other government regulation authorities. *Customer needs*, in turn, are any quality, attribute, or trait of a potential design that is desired by customers. Needs include both "must haves" and "wants". The end customer (user) needs are primary customer needs and usually have the highest importance weighting.

Customer needs that are expressed by actual customers, not imagined by the design team, marketing, or other entities, are often referred to as the "voice of the customer." The voice of the customer is best heard through direct interaction with real customers. This requires that the team conduct individual customer interviews, observe customers in action in the use environment, and when appropriate, gain first-hand experience with the use environment themselves.

Typically, the end user benefits most from the design's functionality, appearance, ease of use, reliability, long life, and safety, but other types of customers such as installers and service technicians also benefit. To be acceptable to end users, the design must meet well defined and clearly understood customer needs in ways that delight and ensure sustainable satisfaction. For other customers such as distributors, middle-men, and so forth, the design should enhance their business, reduce their cost, and contribute to their overall success.

Customer needs are expressed in the "language of the customer". The *design specification* translates the "voice of the customer" into the "voice of the engineer" by defining in precise, measurable terms *what* the design is to do expressed in the language of the engineer. Specifications do not tell the team how to address customer needs, rather they represent an unambiguous agreement on what the design must achieve to satisfy customer needs. In some design situations, it may be necessary to first establish target specifications and then set final specifications later after more is known about the design. The design specification is a formal statement of functional requirements expressed in measurable terms and accompanied by specified limits and target values.

Design Descriptions

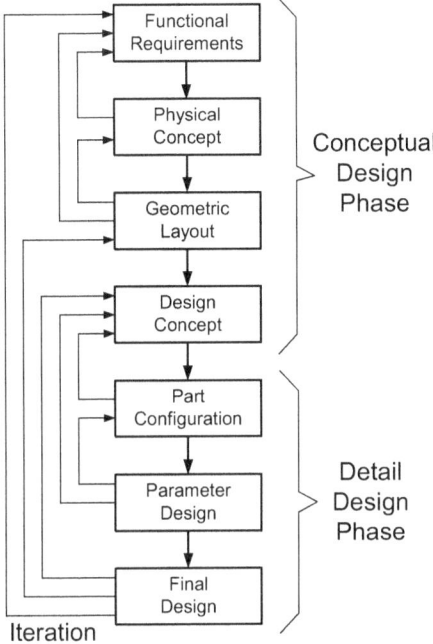

Figure 4.1 Iterative model of the uncertainty reduction process.

The Form Description

The form description of the design is the "how" of the design. It embodies the "conceptual design" and the "detail design" of the designed entity. The *conceptual design* includes the working principle, the design architecture, and the geometric arrangement and configuration of individual parts and components. The *detail design* specifies dimensions, tolerances, surface finishes, and the many other details that are required to complete the form description. Creating the conceptual and detail design involves a process of "uncertainty reduction" during which the design progresses from the abstract to the concrete. This uncertainty reduction process is depicted in Fig. 4.1. As shown, the process begins with functional requirements of the design. Decisions regarding the physical principles upon which the design is to function and the embodiment of these principles into a geometric arrangement of parts is then made. These conceptual decisions define a preliminary initial design which is usually referred to as the "design concept" or "design layout" If the design concept is found to be acceptable, it is then iteratively refined, improved, and "fleshed-out" until the definitive final form description of the design is complete.

Conceptual Design Phase

As shown in Fig. 4.1, the process of creating the conceptual design can be separated into three critical decision activities:

1. **Functional Requirements:** <u>What</u> exactly is the design suppose to do? What are the required functions it must provide or perform?
2. **Physical Concept:** <u>How</u> the required functions are to be fulfilled?
3. **Geometric layout:** <u>How</u> the physical concept is to be divided into a spatial arrangement of separate parts?

The functional requirements, physical concept and geometric layout are developed and merged in the conceptual design phase to define the design concept. This is the critical stage in the design where DFM can have the most beneficial impact on the outcome of the design.

Functional Requirements

Functional requirements (or FR's for short) are concise statements of the functions that the design must provide or perform to successfully implement the functional description of the design. FR's are determined by the *problem definition*, which is the process of going from a primitive recognition of need to a clear and exact understanding of the design problem to be solved. Problem definition is the decisive first step toward good design.

Illustrative Example 4.1: Consider the problem faced by a certain TV station. The problem is that ice forms on the antenna tower during inclement winter weather and subsequently falls off, causing harm to people and damage to automobiles parked below. Concerned about this, the station manager approached a design consultant with the following problem definition: *how to prevent ice from forming on the TV tower*. Although solutions to this problem statement such as installing costly heating elements on the tower structural members could easily be imagined, the consultant, experienced in the need for accurate problem definition, asked the following questions:

- What would happen if ice did form?
- What harm would such formation do?

Following this line of reasoning, a more appropriate problem definition was formulated: *how to prevent ice that forms on the TV tower from doing harm or damage to people and equipment below the tower*. This second problem definition led to an obvious and cost-effective solution that was quickly accepted: build a shed-like structure to protect against falling ice. Not only was this solution clearly superior, it also avoided the costly and time-consuming design and development effort that would have been required by solutions to the first, less accurate problem definition. This example clearly illustrates the importance of making sure the right problem is being solved.

"Preventing harm or damage" is the main FR of the TV tower example. FR's are statements of the design problems to be solved. For example, problems to be solved when designing a plastic beverage container include "contain axial and radial pressure", "withstand moderate impact when dropped", "allow stacking on top of each other", and "minimize the use of plastic". Each of these problems is a FR. FR's may be accompanied with specifications of physical quantities and ranges (e.g., radial pressure in the range of 1 to 3 psi) and are often described verbally such as "enclose", "support", "separate", and "control". In addition, FR's may be distinguished as main, critical, and auxiliary. *Main* FR's pertain to the overall task to be performed. *Critical* FR's are a subset of the main FR's that are of critical importance because they offer the most opportunity for innovation, or perhaps because they are associated with a trouble area. *Auxiliary* FR's, on the other hand, have a supportive or complementary character and generally arise because of the physical concept or geometric layout that is selected.

Function Decomposition Schematic

Functional requirements are determined by decomposing the overall design problem into simpler sub-problems. This is done by creating a diagram termed a *function decomposition schematic* that shows the relationship between sub-problems and the flow of material, energy (or force), and information involved. To create the diagram, first represent the overall design problem as a black box with input and output material flow (double solid line), energy flow (single solid line), and information flow (dashed line). Divide the black box into sub-problems to create a more specific description of what the design must do to solve the overall problem. Each sub-problem is an FR. The function decomposition schematic for a barbeque grill shown in Fig. 4.2 is an example.

In many cases, some of the FR's will have already been assigned to specific components or subassemblies. The schematic should reflect the best understanding of the state of the design, but it does not have to contain every detail. To be manageable, the number of FR's included in the schematic should probably be limited to 30 or less.

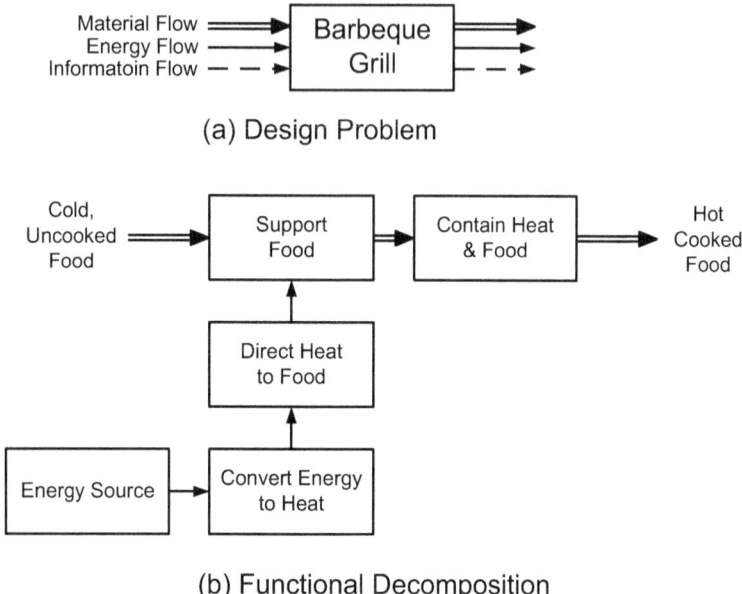

Figure 4.2 Functional decomposition of barbeque design problem.

Physical Concept

The *physical concept* includes (1) the physical principle by which the design achieves its overall function and (2) a physical description of the design embodiment. In other words, the physical concept represents an understanding of exactly "how" the design works and fulfills the functional requirements specified for the design. Developing the physical concept can be a huge accomplishment. It requires that the design team enter the real world of physical principles, materials, how things work, and how things work together. The choice of physical concept adds an immense amount of information to the design, and greatly reduces uncertainty about the design direction. The opportunity for making such significant choices in subsequent stages of the design process rarely occurs. And, once the design is released, changing the physical concept becomes extremely difficult.

The physical concept is decided in many ways depending on the design situation. If the goal is to cost reduce an existing design, then the physical concept is, for all practical purposes, already determined. If, on the other hand, the design is brand new, then identifying the best possible physical concept is of paramount importance. The following examples illustrate alternative approaches for creating a physical concept.

Design Descriptions

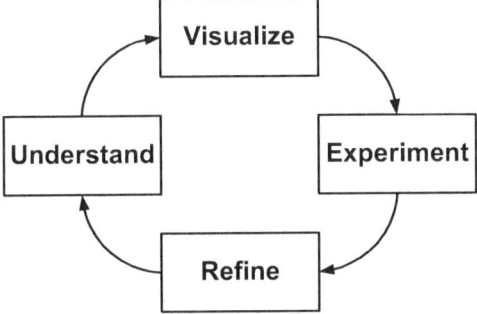

Figure 4.3 The customer-focused design process.

Illustrative Example 4.2: User-Centered Design

User-centered design seeks to identify and evolve the best physical concept for a design through direct interaction with users and the use environment. A multi-step iterative process for understanding user needs, visualizing physical concepts that satisfy these needs, and obtaining user feedback to refine and validate design decisions made along the way is used (see Fig. 4.3).

1. Formulate problem statement
2. **Understand** customer needs and environment.
3. **Visualize** possible design solutions and approaches.
4. **Experiment** by testing promising ideas with end users.
5. **Refine** based on feedback; combine best features into one or more different solution ideas or approaches.
6. Repeat process as necessary.

Understand: Using the problem statement as a guide, understand user needs by interviewing users and observing the use environment. Meet with users to identify and appreciate problems from the user perspective. In addition, carefully observe the use environment, either passively or actively, to discover details that even users are unaware of. Observation can be done passively, such as watching users in action from a distance, or it can be done actively by working side-by-side with a user. Active observation has the great advantage of allowing team members to gain hands-on experience with the use environment and with using existing design solutions. This is especially true in manufacturing, where observing and, when possible and appropriate, actually working on a production line can provide amazing

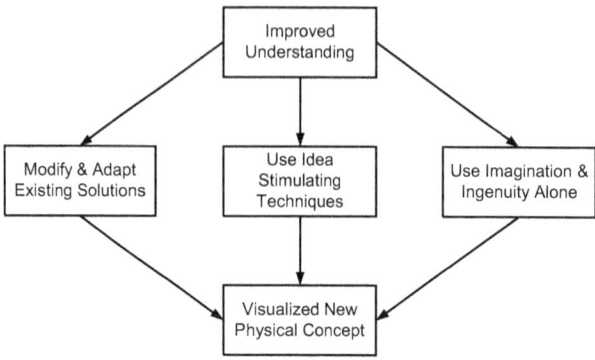

Figure 4.4 Creativity stimulating techniques.

insights and improved understanding of manufacturability issues and challenges. Observations should be carefully documented by taking notes, making audio and/or video recordings, and taking still photographs. Photographs and videos teach a lot about user needs. For example, by displaying a large number of photographs taken of people transporting food coolers at beaches, picnic grounds, tail-gating parties, etc., one design team saw the design solution (mount wheels on the cooler) materialize before their eyes as they viewed the collage of pictures.

Visualize: Imagine possible design solutions (see Fig. 4.4). Use creativity stimulating techniques such as brainstorming. Reverse engineer existing solutions and imagine how they could be improved. It is not unusual for the simple process of experiencing use environment, observing users in action, and participating first-hand with users to stimulate design ideas and solutions by imagination and ingenuity alone. Select among the many ideas generated by winnowing obviously impractical ideas, grouping similar ideas, and combining the best features from the different groups.

Experiment: Experiment with the various ideas by building simple models and mock-ups. Narrow the ideas to the best concepts and test with users. A good way to do this is to construct several models (in practice, 3 is a good number) and include different mixes of promising ideas and innovations in each model. Take the models to the field or factory floor and let customers experience them. Keep track of likes and dislikes for each model. By doing this, the team gets an idea of what design features and innovations are important to users and hopefully avoids making incorrect assumptions about what they think users would like.

Design Descriptions 61

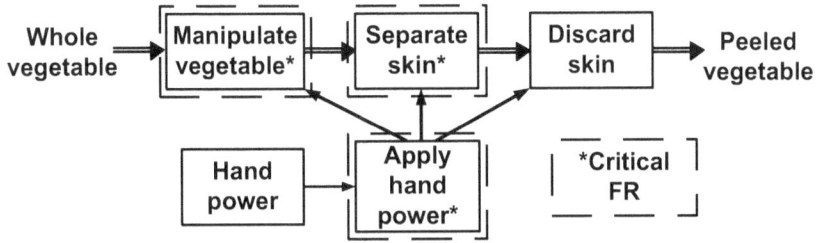

Figure 4.5 Functional decomposition schematic for vegetable peeler.

Refine: Based on feedback and insights gained from the experiment step, define the definitive physical concept by combining the features and innovations that customers like and eliminating those they don't. Take this back to the field and verify the design.

Repeat as Necessary: It may be necessary to repeat the process two or more times. The first iteration establishes a design direction and the follow-on iterations add detail.

Illustrative Example 4.3: The Logical Building Block Method

To see how this simple idea stimulating technique works, imagine that a "better hand powered vegetable peeler" is to be designed. A function decomposition schematic is shown in Fig. 4.5. FR's deemed critical are noted as shown. Sub-solutions for each critical FR are listed in Table 4.1. Alternative designs generated by considering different combinations of sub-solutions are listed in Table 4.2.

Table 4.1 List of possible sub-solutions for each critical FR.

FR1: Manipulate Vegetable	FR2: Separate Skin	FR3: Apply Hand Power
Hold in hand	Knife-blade	Reciprocate
Mount in stand	Wire	Rotate crank
Rotate on mandrel	Abrade	Rock back & forth
Tumble in bowl	Tool	Twist

Table 4.2 Alternative physical concepts based on Table 4.1 sub-solutions.

Sub-solution Combination	Alternative Physical Concept
Tumble in bowl + abrade + rotate crank	Sandpaper lined rotating bowl
Hold in hand + knife blade + reciprocate	Conventional peeler
Rotate on mandrel + tool + rotate crank	Lathe like device.

Geometric Layout

From a DFM standpoint, the choice of geometric layout is one of the most far-reaching and significant design decisions that is made during the course of product development. It often determines the part count, the assembly sequence and precedence constraints, ease of assembly, ease of serviceability, amount of quality risk, standardization possibilities, and much more.

To determine the geometric layout of a design, disassemble it, keeping track of each part as it is removed, and then reassemble the parts in the reverse order. The number of parts, their geometric arrangement, the inter-relationships and interfaces between them, and the way they assemble and integrate to form the completed design define the geometric layout. By taking a design apart after it has been designed and manufactured, the design is essentially being "decomposed" into individual parts. For this reason, the terms "geometric layout" and "part decomposition" both refer to the same thing. In this book, the term "geometric layout" is preferred. In other books and articles that I have authored, "part decomposition" has been used.

The *geometric layout* defines the spatial arrangement of the designed parts, the standard components, and sub-assemblies together with the way they are integrated together by the assembly to create the finished design. The layout may be created in two or three dimensions as a sketch, a computer solid model, or even possibly as a 3-D physical model constructed from cardboard, foam, or other found materials. Typically, the geometric layout begins as a very preliminary concept because, at most, only key dimensions and relationships have been specified and the shape and size of components are either estimates based on a mixture of guesswork and calculation or they are represented by simple "placeholder" geometries. The final definitive geometric layout results from the complex interplay between (1) the material and process selection and (2) the assembly structure.

The *material and process selection* involves the choice of the basic material class (e.g., metal, plastic, wood) and type of process (e.g., machining, sheet metal forming) that is chosen for each designed component. In many mature industries, material and process are pre-determined or implied. For example, the exterior components of a typical passenger car are likely to be formed from sheet metal. In a totally new product development, the options can be much broader. Depending on production quantity and other considerations, an enclosure for a small electronic device (e.g., computer or cell phone) could be CNC machined from solid metal, formed using plastic and a 3-D printing process, formed out of sheet metal, designed as an aluminum casting, or molded using an engineered plastic and a polymer processing method such as injection molding.

The *assembly structure* of a design encompasses the way in which the designed parts, standard components, and sub-assemblies are supported, oriented, located, joined, and integrated together to form the whole. Most assembly structures utilize a chassis or frame on which components are mounted. Frames can be oriented vertically (e.g., bicycle or motorcycle frame), horizontally (e.g., rail-type automotive frame or electronics chassis), or designed as 3-D skeletons (e.g., space frame, unibody, card cage, etc.). Parts can be located, oriented, and joined to the frame or base part in different ways such as the use of nesting features or the use of a multitude of different joining processes. As discussed in Chapter 10, information content can be reduced in many ways by "designing for assembly".

The geometric layout is pivotal from a DFM perspective because of the far-reaching consequences it entails. In addition to establishing the work content, the geometric layout determines the assembly sequence, the precedence constraints, and the minimum rational work elements into which the work content can be divided (see the section on production lines in Chapter 3). Ambiguity and information content are greatly reduced when factors such as these are considered early in the conceptual design.

Design Concept

The physical concept and geometric layout together define the *design concept*. It is the combination of physical concept and geometric layout that determines, often synergistically, important design characteristics such as manufactured quality, ease of assembly and service, and manufacturing cycle times. It is for this reason that design for manufacture focuses so intently on identifying the best possible initial design concept, which as a rule, is the one having the fewest, easy to make and assemble parts. For many designs, the geometric layout can be as important as or even more important than the physical concept. The geometric layout determines much of the direct and indirect cost of the product. It determines the number and complexity of the designed parts, which in turn, influences tooling cost, tolerance stack-up, smoothness of force-flow, and numerous other engineering considerations. In addition, as alluded to above, the geometrical layout determines the assembly processes that must be used as well as the sequence in which the parts are assembled. This in turn determines the number of workstations and material flows on the production line.

The design concept establishes the "intrinsic cost" of the design. *Intrinsic cost* is the lowest imaginable "total cost" that is possible when the final design perfectly satisfies all the functional requirements of the design, uses the theoretical minimum number of ideally shaped parts, and is perfectly configured for minimal fabrication and assembly cycle time. In general, intrinsic cost is unachievable practically. It is useful, however, as a criterion for design concept selection because the design concept having the lowest intrinsic cost has the greatest potential for achieving the lowest total cost through continuous improvement over time.

To illustrate, consider the design of an air compressor to meet specified pressure and air-flow requirements. Several alternative design concepts are possible, each utilizing a different physical concept and geometric layout. For example, the overall function of raising air pressure could be accomplished by using a reciprocating piston, a scroll, or possibly a rotary vane. A reciprocating piston air compressor works like a car engine. It uses a crankshaft rotated by a drive motor to reciprocate a piston back and forth, compressing air that is trapped between the piston and the cylinder head. A scroll air compressor uses two interleaving scrolls, each shaped in the form of an involute curve (Archimedean spiral) to compress the pocket of air that is trapped between them. Rotary vane compressors use a vane that reciprocates radially in an eccentrically mounted rotor to compress air that is trapped between the vane, rotor, and housing.

Which design concept should be chosen? Both the scroll and rotary vane compressors require fewer parts compared to the reciprocating piston design. And compared to the scroll, the vane compressor is composed of parts that are easier to manufacture and have fewer critical tolerances. Therefore, assuming there is no mitigating functional, business, marketing, or other constraints, the vane compressor is likely to have the lowest intrinsic cost thus providing either greater profit or enhanced pricing flexibility.

In some designs, the physical concept and geometric layout are, by their very nature, tightly coupled. In these cases, the physical concept usually dictates the geometric layout, as illustrated by the air compressor example. In other situations, such as the design of an electronic assembly, coupling is weak, and many alternative geometric layout choices are possible for a given physical concept. Most design situations lie somewhere between these extremes. In all cases, the aim of DFM is to approach intrinsic cost as closely as possible.

Detail Design

Detail design (see Fig. 4.1) involves developing a configuration design and parametric design for each designed component that implements the design concept. *Configuration design* includes determining the size, shape, and detail features of the designed components, while *parametric design* involves assigning specific material properties, dimensions, and tolerances. During detail design, the design changes and evolves as questions are answered and uncertainties resolved. The result of the detail design stage is a definitive final design that includes all the information required to fabricate and assemble the parts.

Traditionally, detail design decisions have focused primarily on reducing piece-part material and manufacturing cost. Too much emphasis on piece-part cost can drive the design toward geometric layouts that are composed of many low-cost parts, thus minimizing direct cost rather than total cost. Because DFM seeks to minimize total cost, in some design situations, piece-part cost may increase. This is when it is extremely important to understand that piece-part cost is just one component of total cost. With DFM, piece-part cost may be higher but, because the part count is greatly reduced, total cost will invariably be lower. A well-conceived geometric layout not only allows and facilitates piece-part designs that are easy to tool and fabricate, it also reduces assembly cost, cost due to quality risk, cost of non-value-added manufacturing activities such as material handling, and it leverages the benefits of standardization and platform designs. Often, a well-formulated geometric layout facilitates economies of scope and scale that would be impossible if each designed part is individually cost reduced. The key is to see the big picture rather than focusing on individual details. The importance of a total cost focus is a constant reoccurring DFM issue. This is largely because direct cost is relatively easy to estimate and, at the same time, total cost is next to impossible to estimate in any realistic way. Avoiding the direct verses total cost issue is a primary concern in Chapter 6.

The Fabrication Description

Better known as the "manufacturing plan", the *fabrication description* specifies how the design is to be manufactured. Historically, creating this plan has been the responsibility of manufacturing engineering. It usually consists of two major components: (1) the process plan, and (2) the production plan. The *process plan* specifies the steps and operations required to fabricate and assemble the design. Depending on the nature of each operation, the plan will also include details concerning the machines, tools, process parameters, setups, and other information needed to perform each operation. Identifying the best process plan often requires analysis of alternative approaches and trade-offs that are driven by considerations of available capabilities and resources, investment requirements, and so forth.

The *production plan* spells out how the process plan is to be implemented in the firm's production facility. This plan can range from a simple "routing sheet" that specifies how work units move through the production facility and what operations are to be performed at each stop along the way, to a complex and comprehensive plan for producing a major product such as an automobile, airplane, or ship. The production plan typically includes diverse considerations such as production line layout, workstation design, material handling and storage, and scheduling and production control. For large projects, system integrators and other contractors may be utilized to provide design, build, installation, and maintenance services. Similarly, outside tool design and manufacturing specialist may be used. This is often the case for parts such as castings and powder metallurgy products that require special knowledge and expertise.

Illustrative Example 4.4

To illustrate how the fabrication description (manufacturing plan) is created, consider the part shown in Fig. 4.6, which is to be machined from 18-foot lengths of 2-inch diameter bar stock. Available machines of suitable capability include:

- A screw machine that machines in one direction along the part length and at the free end ($26/hr).
- A CNC lathe capable of automatically flipping and clamping the part to allow machining in both directions ($51/hr).
- A milling machine ($23/hr).
- A centerless grinding machine ($38/hr).
- A cut-off saw ($12/hr).

The following three-step process is typically used for developing a manufacturing plan for the part.

Design Descriptions 67

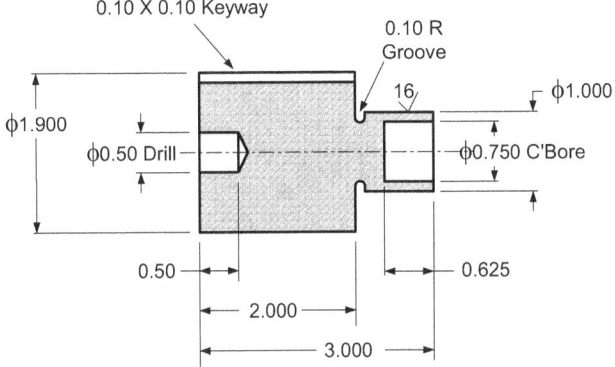

Notes:
1. Material: CRS (cold-rolled steel)
2. All dimensions are in inches.
3. Diameters having "3-place" accuracy must be finished turned.
4. The 0.10-inch radius groove can be cut using a form cutter.

Figure 4.6 Production drawing of a part that is to be made in the firm's production machine shop.

1. Develop a tentative process plan for machining the part. Each process plan specifies the machines and machine set-ups to be used. For each set-up, the plan further specifies the sequence in which the operations are to be performed. In developing the alternative plans, the manufacturing engineer accounts for special requirements specified by the drawing. For example, Note 3 (see Fig. 4.6) requires that all external cylindrical surfaces be finish turned. The "16" surface finish symbol placed on the 1.000-inch outside diameter requires that this surface be finish ground.

2. To choose the best process plan, total machining time per part and cost per part are estimated using recommended cutting conditions and approximate operation times. These results are then compared, and the most economic plan is chosen.

3. Once an acceptable process plan has been developed, the manufacturing plan is developed by generating a "routing sheet" or "traveler" that specifies all needed detail information required to process the parts through the plant. This generally includes batch size, details specifying the sequence of machines and the setups for each machine, the sequence and types of operations to be performed for each setup, the processing parameters for each operation (speeds and feeds), and so forth.

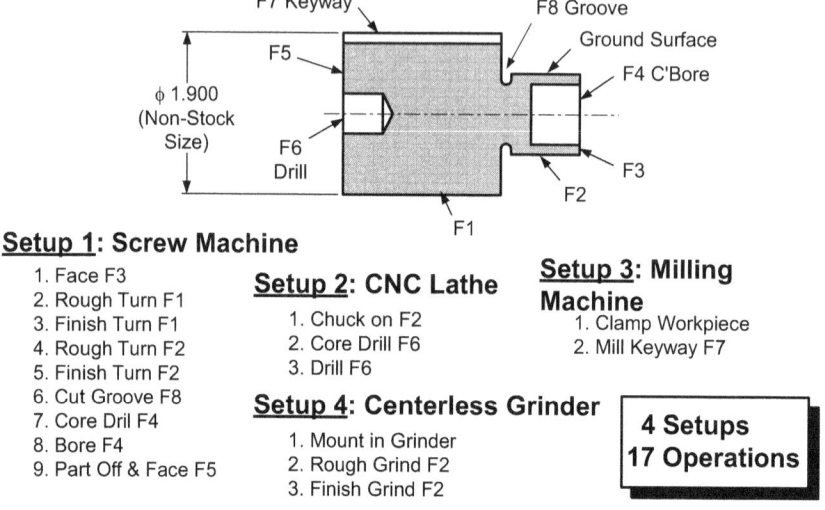

Setup 1: Screw Machine
1. Face F3
2. Rough Turn F1
3. Finish Turn F1
4. Rough Turn F2
5. Finish Turn F2
6. Cut Groove F8
7. Core Dril F4
8. Bore F4
9. Part Off & Face F5

Setup 2: CNC Lathe
1. Chuck on F2
2. Core Drill F6
3. Drill F6

Setup 4: Centerless Grinder
1. Mount in Grinder
2. Rough Grind F2
3. Finish Grind F2

Setup 3: Milling Machine
1. Clamp Workpiece
2. Mill Keyway F7

4 Setups
17 Operations

Figure 4.7 Process plan for the part of Fig. 4.6.

The machining time and cost per part for two alternative process plans are shown in Table 4.3. These values were calculated using best practice cutting data available to the manufacturing engineer. The bar stock must be cut-to-length either by using the cut-off saw or by "parting-off" using the screw machine. The first and second alternative process plans in Table 4.3 utilize the cut-off saw and screw machine, respectively. As shown in Table 4.3, the second alternative provides the lowest machining time and cost per part. A "process plan" for this alternative is shown in Fig. 4.7. This plan requires a total of 4 setups and 17 operations and results in an estimated cost per part of $2.84.

Table 4.3 Machining time and cost per part for alternative process plans.

Alternative Process Plan	Machining Time per Part	Manufacturing Cost per Part
1. Saw, CNC, Mill, Grind	7.93 min	$5.20
2. Screw Machine, CNC, Mill, Grind	6.20 min	$2.84

Function-Form-Fabrication Interactions

Example 4.4 clearly illustrates the interplay that occurs between the function, form, and fabrication descriptions of the design. In looking at this example, we see that the key cost drivers are the number of setups, the number of machining operations per setup, and the cost of the machine required to perform the operations. Setups drive cost because they take time, waste material, and add quality risk. Each setup requires that the machine be taken "off-line" thereby curtailing its productivity. In addition, setups produce scrap and waste because, in most cases, the machine must be run, the product checked, and machine parameters readjusted, until acceptable product is reliably produced. Each machining operation performed per setup drives cost because it requires tools and tool changes, represents a quality risk, and increases cycle time. The production machinery drives cost because they are unavailable for other purposes, require power and floor space, and add to the amount and type of material handling required.

The interaction occurs because there is little that manufacturing engineering can do about these cost drivers since they are literally "specified" by the function and form descriptions of the design. Because the design is "frozen" when it is released to manufacturing, about all that manufacturing engineering can do is to experiment with speeds and feeds to reduce machining cycle time. Beyond this, the only other cost reduction opportunity is to upgrade the firm's flexible manufacturing capability allowing more to be done with fewer setups. But this requires long-term planning and investment decisions that would not be made just to reduce the cost of one part.

Clearly, the real opportunity for true cost reduction is to avoid these extra costs in the first place by including consideration of manufacturing needs and constraints early in the development of the functional and form descriptions of the design. When this is done as it is in the DFM approach, unneeded setups, processing operations, material handling requirements, and other unnecessary manufacturing steps can be eliminated. As a result, the extra cost required by these needs is never incurred. And, perhaps more importantly, they never become production problems or cost drains that continue for the life of the design. In Chapter 5, we return to Example 4.4 to illustrate the power of the DFM approach.

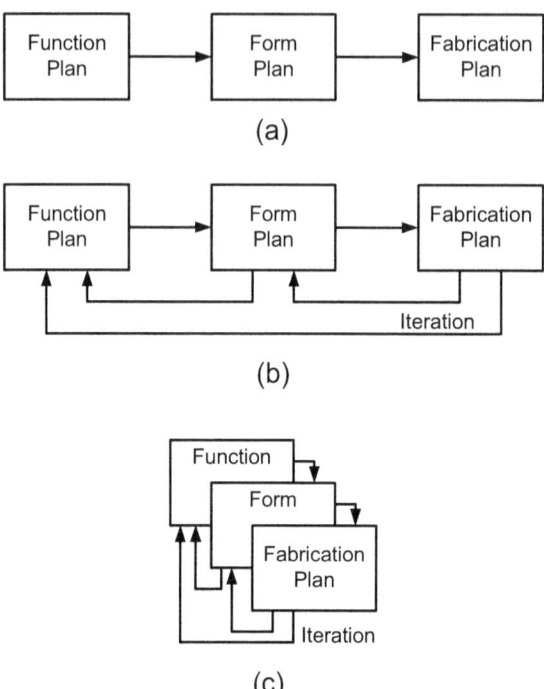

Figure 4.10 Product design process: (a) linear and sequential; (b) non-linear and iterative; (c) overlapping and coordinated.

The Overlapping Nature of the Design Descriptions

Traditionally, the three design descriptions have been developed in a linear fashion starting with function, then form, and finishing with fabrication as shown in Fig. 4.10a. The functional description defines "what" the design is to achieve, the form description defines "how" the "what" is to be achieved, and the fabrication description defines "how" the form description is to be achieved. The two "how's" in this chain imply that the design process is not as linear and sequential as the traditional approach seems to suggest.

The interplay between "what" and "how" is far more complicated than it may at first appear because of the design iteration that it requires. As shown in Fig. 4.10b, this interplay involves a non-linear iterative "give and take" process of accommodation and trade-off between function, form, and fabrication. Because of the close coupling between each of the design descriptions, if not carefully managed, this interplay can result in undesirable mismatches between each description that invariably increase total cost and time while also degrading total quality of the design.

Iteration allows the design to be continuously improved and optimized as better and more complete design information about each design description becomes available. What constitutes "more complete information" can vary depending on circumstance. For example, more complete information could come in the form of improved understanding gained from a prototype test or revealed through analysis or innovation; it could be a better definition of customer or market needs or a change in those needs; or it could involve a new material or emerging manufacturing technology or some other new innovation.

Accompanying iteration are local design changes made to improve the design. These local changes generally propagate in a "ripple effect" throughout the design because they require that each part of the design affected by the change be reexamined. "Engineering change", caused by both the design change itself and the ripple effect it produces, is the inevitable result. In the early phases of a design project, engineering changes are easily handled because the design is fluid, hardware is still remote, few people are involved, and constraints and interactions have not become tight.

In the later stages of the project, engineering changes become much more difficult and costlier to handle; many engineers, designers, and technicians are committed, several components of the manufacturing system are usually actively involved, and much has been designed and irreversibly fixed. Because of the ripple effect, an ill-chosen solution to even a relatively minor problem can put the whole project in jeopardy. For these reasons, the range of solutions to a problem discovered late in a project are severely limited, and even minor changes are likely to result in both undesirable deviations from the original design intent and in suboptimal design.

Cost of engineering changes made late in the project can be devastating. Direct cost is high because of time delays and the large amount of personnel involved. Indirect costs increase because of the suboptimal solutions that must be accepted to contain direct costs of the change. Suboptimal solutions generally lead to higher manufacturing cost as well as a degradation in manufactured quality and manufacturing productivity. These costs, of course, continue for the life of the design.

When engineering changes occur after production release or originate because of production difficulties (design situation #4, Chapter 1), the cost incurred include additional indirect costs such as scrap parts, wasted material, and idle machinery. After sales, engineering changes usually trigger high warranty or service costs, which are also indirect costs that must be accounted for. Intangible costs, such as loss of reputation and declining workforce morale also represents very real business costs.

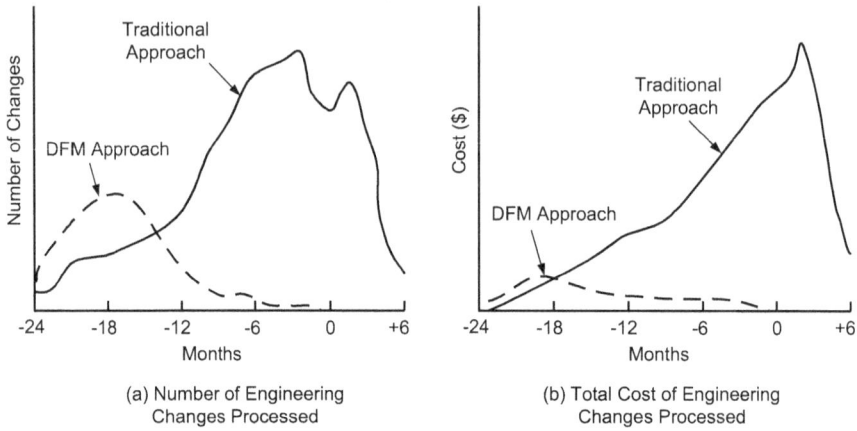

(a) Number of Engineering Changes Processed

(b) Total Cost of Engineering Changes Processed

Figure 4.11 Comparison between a design project that followed a conventional approach and a similar one that utilized the DFM approach.

Design for manufacture seeks to move engineering change to the early stages of design and thereby avoid these undesirable consequences. This is done by making function, form, and fabrication information available early in the design process when it is needed. The goal is to "do it right the first time". Achieving this goal requires that all three design descriptions be developed in an overlapping, coordinated, essentially parallel process as shown in Fig. 4.10c.

The benefits of the DFM approach are illustrated in Fig. 4.11. By moving engineering change into the early stages of design, the DFM approach dramatically reduces the amount of engineering change required as well as the resultant cost incurred by making the changes. In this illustration, the cost of change is estimated using the "law of 10s", which assumes that the cost of an engineering change increases by a factor of 10 with each subsequent stage in the product realization process.

In Fig. 4.11, "0 months" represents production of "job 1", when the first production product comes off the line. Note the dip in engineering changes for the conventional approach leading up to the "job 1" event. This is the result of a "freeze" being placed on the design to enable the product launch. In some cases, the product launch date is driven by the project schedule, which initially is arbitrarily chosen but ends up being the "tail that wags the dog". The total cost consequences of the "schedule driven" mentality that dominates thinking in some companies can be devastating. Shifting the launch of "Job 1" from a schedule driven date to a "production readiness" criteria is often an important benefit of the DFM approach.

Summary of Key Concepts

- ➤ A complete design consists of three design descriptions: function, form, and fabrication.
- ➤ Each design description is critical to the design's success.
- ➤ From a DFM perspective, the geometric layout is the most consequential design choice that is made during the design project.
- ➤ Opportunities exist for optimizing the geometrical layout by considering all three design descriptions concurrently.

Chapter 5
The DFM Approach

Design for manufacture is concerned with achieving three key design goals:

1. Identify design concepts that optimally satisfy customer needs and that are also inherently easy to manufacture.
2. Focus on component design for ease of assembly and fabrication.
3. Coordinate the design and manufacturing concepts to achieve optimal matching of needs and requirements.

The DFM approach presented in this chapter is a systematic and disciplined methodology that is precisely aimed at achieving these goals. It is based on a firm belief that DFM is achieved by maximizing total design value. The approach is predicated on five *DFM principles*:

1. **Concurrent Engineering Principle:** The design concept and the manufacturing plan should be developed in a concurrent manner as an "integrated product/process" design.
2. **Team Principle:** A team approach that represents the view of all stakeholders in the design should be utilized.
3. **Good Design Principle:** All design decisions, both large and small, should help maximize total design value by adhering to the rules of good design, the design axioms, and the many DFM guidelines that have been derived from design and manufacturing experience.
4. **Structured Methods Principle:** Structured methods should be used when appropriate to provide insight and to stimulate creativity and innovation.
5. **Stretch Objectives Principle:** Stretch objectives that push the envelope and challenge the design team should be used when appropriate to, again, stimulate creativity and innovation.

The DFM Approach

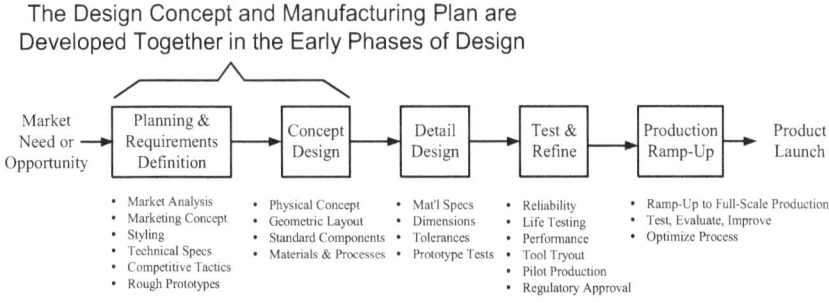

Figure 5.1 Phases of the product realization process.

Concurrent Engineering Principle

A manufacturing system comprises numerous distinct aspects, processes, and stages that interact in highly unpredictable ways. For this reason, many of the design decisions made during design development negatively impact the production process in hard to foresee direct and indirect ways. Imagine, for example, a family of products that has, over time, been differentiated into a "hard" variety of different sizes that makes each model incompatible with the others. The unintended consequence is added manufacturing complexity including the need for separate production lines, model specific tooling and fixturing, and numerous other undesirable duplications of effort. Had the needs of the production system been considered early in the design of the original product, it is likely that many of these consequences could have been avoided.

As examples such as this clearly demonstrate, the result of not considering "downstream" manufacturing and support needs early in the design process is suboptimal design of both the product and the manufacturing process. Experience over the past thirty or more years has unquestionably shown that, to avoid problems such as these, the product design and associated manufacturing plan need to be developed concurrently as an integrated product/process design, early in the product and process development cycle (see Fig. 5.1), before concept and equipment decisions have been made. In the DFM approach, the goal is to create the design concept and the manufacturing plan concurrently so that the needs of both the form description and the fabrication description of the design are considered simultaneously. This idea is variously known as "concurrent engineering" or "simultaneous engineering", depending on the company and industry involved and has been standard practice in many companies for many years. It is an essential feature of the DFM approach.

Team Approach Principle

A design engineer who takes manufacturability into account when creating a design will no doubt create a more cost effective and producible design. The reality is, however, that even if the design engineer has an unusually comprehensive understanding of manufacturing requirements, processes, and operations, a single individual's knowledge can't compare to the potential for DFM improvement that is possible when a diverse set of perspectives and expertise is brought to bear. In the DFM approach, all stakeholders in the design need to be involved from the very beginning.

In the team approach, interactions between function, form, and fabrication are communicated, coordinated, and controlled in a continuous on-going dialog. Function, form, and fabrication solution proposals are openly shared in order to ensure coordination and to avoid conflicts. When conflicts arise, they are resolved in a cooperative manner. The key to success lies in the "give and take" facilitated by the team approach. Assurance that all technological innovations, manufacturing variables, and other downstream needs have been properly considered can only be accomplished by give and take between several points of view.

The main purpose of the team approach is to ensure that all needed information is readily available as design decisions are made. The team should, therefore, be cross-functional so that all points of view and disciplines are represented. Cross-functional teams enhance design creativity through cross-fertilization of thought processes, behaviors, and functional skills. They also facilitate nonlinear design iterations that bounce between disciplines by ensuring that design decisions are fully informed (see Fig. 5.2) thus avoiding the engineering change, lengthy redesign cycles, and costly "over-the-wall" iterations that typically occur when the design process proceeds in a more linear fashion.

As the project goes through its various phases, different skills and numbers of people are required. To accommodate fluctuating resource needs and team size, the project team can be divided into a core team and an extended team. The *core team* is composed of a small number of individuals with complementary skills who have day-to-day responsibility for the project. Core team members are involved with the project from the start and typically continue until the product is tooled and in production. The *extended team* includes all those who bring needed skills to the project and are involved at different stages of the project or who participate on an as needed basis.

The DFM Approach

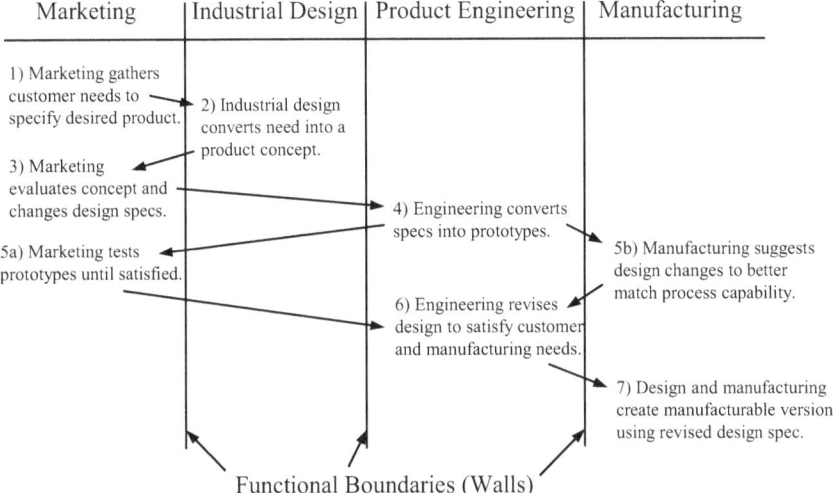

Figure 5.2 Cross-function design teams facilitate nonlinear iterations instead of linear "over-the-wall" iterations.

The choice of team leader is another important consideration. In addition to a strong commitment to DFM, an effective team leader will generally possess the following desirable characteristics (Ettlie and Stoll, 1990):

- Respected both for business judgement and technical expertise.
- Total dedication, to the point of obsession, with the project.
- Charisma that conveys a strong impression to others of knowing the direction to go, welcoming other travelers, and convincing them to follow.
- Stamina to continue with the project until all goals are met.
- Positioned to have authority to cause instant action without having to negotiate, arbitrate, or plead for support.

Good Design Principle

The goal of good design is to maximize total design value. As discussed in Chapter 2, this is done by using the rules of good design and the design axioms to guide every design decision, both large and small. Because the concept of good design is relatively abstract, it is helpful to formulate it in ways that are easily applied to design decisions. This is where detail DFM guidelines come in. *DFM guidelines* implement the design axioms by translating them into specific DFM advice and guidance.

DFM guidelines can be thought of as good design expressed in the practical language of design and manufacture. Like the design axioms, the guidelines are stated as directives that act to stimulate creativity by showing how a design can be improved. And, because the design axioms are fundamental to all design, the DFM guidelines correlate with the wealth of empirical design approaches, techniques, "tricks of the trade", and design tips that have evolved out of design and manufacturing experience. When correctly implemented, the DFM guidelines will result in designs that are inherently easy to manufacture and support. A broad range of DFM guidelines are derived in Chapters 9 through 11. The following short list of randomly selected DFM guidelines illustrate their nature and how they work.

1. Minimize the Total Number of Parts
2. Design Parts for Ease of Fabrication
3. Avoid Separate Fasteners
4. Minimize Assembly Directions
5. Minimize Handling

Almost all DFM guidelines including those listed above work to reduce information content in one way or another. Some are particularly effective. Minimizing the number of individual parts used in a product is a good example. A part that is eliminated costs nothing to design, make, assemble, move, handle, orient, store, purchase, clean, inspect, rework, service, or replace. It never jams or interferes with automation. It never fails, malfunctions, or needs adjustment. It requires no solid model or part number and never generates an engineering change notice (ECN).

In automation applications, separate fasteners are difficult to feed, tend to jam, require monitoring for presence and torque, and require costly fixturing, part feeders, and extra workstations. Avoiding separate fasteners eliminates all this extra automation information and quality risk. The same holds true for assembly directions and material handling. By designing for "top-down" assembly, many degrees of freedom and the information that goes with them are eliminated. Use of part magazines, tube feeders, part strips, palletized trays and kitting techniques help preserve part orientation thereby greatly reducing the information required to manipulate and handle parts during assembly.

The DFM guidelines presented in this book are, for the most part, applicable to broad categories of design for manufacture concerns such as assembly and piece-part design. But the concept of DFM guidelines is not limited to just these. By using the design axioms as a guide, it is possible to derive additional DFM guidelines that are specific to a given type of product or process. Also, close examination of many unwritten understandings and "rules of thumb" will show that they are implementations of the design axioms. For example, reduced information content explains why near net shape manufacturing processes are preferred.

Structured Methods Principle

Structured methods are procedures, techniques, tools and best practices that help guide and facilitate the solving of design problems. They benefit DFM in two important ways. First, structured methods formalize design procedures thereby providing discipline and objectivity while also helping to (1) avoid oversights and other errors that can occur with more informal or haphazard approaches, (2) broaden the scope of the problem-solving activity to ensure that all viable solutions are identified and considered, and (3) document the decision-making process. Second, by making DFM goals more explicit, structured design methods help facilitate good design communication and the team approach. All team members can see and understand what is going on and contribute to the process. Group consensus is quickly achieved and the probability of making the best possible design decision is greatly enhanced.

As an illustration of how structured methods work, consider a simple checklist. By listing key elements or tasks, the checklist formalizes the process and makes it explicit. As items are performed, they are checked off until everything is complete. This ensures that items won't be overlooked or forgotten and frees personnel from the need to carry a multitude of details in their heads. Teamwork is facilitated because the checklist makes needed information available to all team members. Also, individual tasks on the list can be allocated to different team members, saving time and improving efficiency. When the process is complete, the checklist becomes a record of accomplishment for future reference.

Structured methods formalize various procedures of design and make them explicit. They can be used at all phases and levels of design. Using these methods requires extra effort and may appear, on the surface, to be diverting time and attention away from the central task of producing a design. Experience has consistently shown, however, that structured methods save time by bringing logic and rationality to what can be a very open ended, random, and chaotic process. As an added benefit, they force strategic thinking about design and manufacturing.

Use of structured methods requires discipline and focus as well as vision, resource commitment, and long-term management support. In general, expect effectiveness to increase with practice and as experience is gained through repeated use. Don't become discouraged, however, if the first few attempts take longer than expected or yield less than hoped for results. Using structured methods is a process that must be continually improved. A sampling of widely used structured methods includes:

- **Design for Assembly.** The Design for Assembly (DFA) method was developed by G. Boothroyd and P. Dewhurst. Details of the methodology are presented in *Product Design for Assembly* (Boothroyd, 1994). The goal of the method is to increase assembly efficiency, which is defined as the theoretical assembly time divided by the actual assembly time. The "Geometric layout Improvement Method" presented in Chapter 12 is a simplified version of the Boothroyd-Dewhurst DFA method.

- **Design for "X" Methods.** DFX methods seek to provide guidance to the designer to help ensure that parts and products are correctly designed to be compatible with a given manufacturing process or activity. Design for casting, design for machining, design for testing, and design for service are examples as are methods that teach how to design using a company's product or service.

- **Failure Mode and Effects Analysis (FMEA).** FMEA is an important design and manufacturing structured method intended to help prevent failures and defects from occurring and reaching the customer. It provides the design team with a methodical way of studying the causes and effects of failures before the design is finalized. Similarly, it helps manufacturing engineers identify and correct potential manufacturing and/or process failures.

- **Taguchi Method.** The structured method that has come to be known as the Taguchi Method seeks to optimize the design with respect to robustness (see Chapter 13). The method works by helping to relax tight tolerances by reducing sensitivity to hard-to-control variation.

- **Group Technology (GT).** This practice seeks to reduce manufacturing system information content by identifying and exploiting the sameness or similarity of parts based on their geometrical shape and/or similarities in the production process.

For design methods and tools such as these to be used effectively, it is important that everyone in the organization understand the purpose of the methodology and how it is to be used. This is because they require training, discipline, design time, and management commitment which can make their value questionable. Common complaints include: "How is this going to help?" "We haven't got time for this". "I can't afford to dedicate valuable resources to this". To avoid this kind of resistance, a clear, unambiguous vision for the method needs to be emphasized. This includes why the methodology is important, when and how it is to be used, and how the results will be used. Often, structured methods are the most effective when they are made integral to the firm's design process and institutionalized as being "the way we do things".

Stretch Objectives Principle

In applying the DFM approach, it is often useful to set "stretch objectives" early in the design. Examples might include no separate fasteners in final assembly, no reorientations of the build during final assembly, no dangling connectors or loose, unconstrained wires, no tension springs, and so forth. Often, manufacturing and support needs expose opportunities for setting stretch objective. Another way to develop a realistic set of stretch objectives is to carefully list the advantages and disadvantages of the design and manufacturing methods currently being used. A rich source of knowledge for doing this are the factory personnel, assembly operators, line supervisors, and other factory employees who live with the challenge of making the design every day. No one is more knowledgeable about existing design-related production problems or more qualified to offer credible advice about how to correct or avoid a problem. The key is to listen patiently without judging and without expecting design improvement ideas. Factory workers know what is wrong about the design, they may not have plausible ways for improving it.

Another way to set stretch objectives is to reverse-engineer leading competitor products. If a particularly clever or simple design solution to a problem is found, the design team should be challenged to either beat the competitor's solution or use it. Many designers and engineers, when faced with this challenge, find amazingly creative and innovative ways to improve on their competitor's design. When stretch objectives are set in this way, probabilities become high that the resulting product will be superior to the competition in the areas where the stretch objectives are achieved.

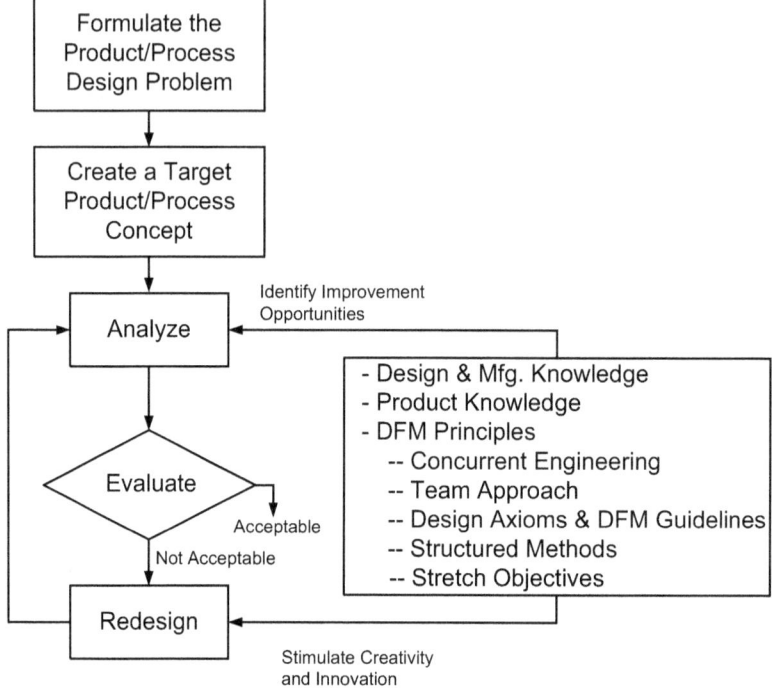

Figure 5.3 The design-analyze-redesign DFM approach.

DFM Approach

An effective DFM approach must ensure that "downstream" manufacturing and support needs are carefully considered. In addition, it must also (1) minimize the chance of forgetting or over-looking an important consideration, (2) be flexible enough to meet the needs of widely differing design situations, and (3) be simple enough to become habitual in its use. The DFM approach presented here is based on these requirements. As illustrated in Fig. 5.3, the approach utilizes a simple design-analyze-redesign process that begins with the formulation of the product/process design problem to be solved. The design problem can involve the design of a part, an assembly, a product, or a system. The key to the DFM approach is the use of the DFM principles (concurrent engineering, team approach, good design, structured methods, and stretch objectives) to guide and inform each step in the process. This helps to ensure that every detail of the design has been considered and optimized from a DFM perspective.

As illustrated in Figure 5.3, the DFM approach involves the following basic steps: formulate the design problem, create an initial target product/process concept, evaluate the current concept, decide acceptability, and create a redesign if necessary. The process is repeated iteratively until an acceptable design is achieved. Each of these steps is discussed below in greater detail.

Formulate the Product/Process Design Problem. This step involves, first of all, identifying the type of problem to be solved and putting together a design team that has the design and manufacturing knowledge and expertise needed. Then, to complete the formulation, initial information that describes and poses the problem is obtained, refined, and checked for accuracy and completeness (since it will seldom be complete and may sometimes even be inaccurate). Finally, the problem is formulated into a standard structure for solution by known methods and available tools.

Create Target Product/Process Concept. The target product/process design is a combination of design concept (working principle and geometric layout) and manufacturing plan (see Chapter 4). Creating an initial product/process design generally involves the use of experience, creativity, physical principles, qualitative reasoning, and much more. Focus is typically on satisfying the design specification including consideration of functional requirements, customer needs, cost targets, and other design goals. In general, the initial manufacturing plan should be an ideal target toward which to work. The initial design concept can be either an existing design or a new concept depending on the circumstances of the design problem.

Analyze. The purpose of this step is to identify potential manufacturability problems and to discover improvement opportunities. The analysis is performed by the design team using their experience and product knowledge together with the design axioms and DFM guidelines as a guide. In addition, analysis methods and procedures may be used to predict performance, estimate cost, and assess ease of manufacture. For example, structured methods such as the "geometric layout improvement method" (see Chapter 12) may be employed, either to gain useful insight or to identify unanticipated weaknesses or potential issues.

Decide Acceptability. The candidate product/process concept is acceptable when it is judged to be "good enough" to move on to the next stage of the realization process. Deciding to continue to invest time and money in efforts to make improvements is a constant dilemma. Fortunately, convergence to an acceptable design solution is generally very rapid when needed design information is available and the design axioms and DFM guidelines are used as the basis for design decisions.

Redesign. When the current design solution is judged to be not good enough, redesign using the design axioms and DFM guidelines together with the results of previous analysis and evaluations to improve the design. The power of the DFM approach lies in this redesign step. When performing this step, the design team visualizes and/or imagines how downstream assembly, part processing, service and maintenance, and other lifecycle operations and considerations are to be performed or otherwise accounted for. The focus is on minimizing work content, optimizing operation sequence, minimizing assembly precedence constraints, and providing "designed-in" features that facilitate, simplify, and ease "downstream" processes. By creatively applying the DFM principles, innovative product/process solutions are found to potential quality, manufacturing and lifecycle support problems. The ultimate result is an optimal geometric layout combined with a coordinated manufacturing concept that helps maximize total design value.

The initial target product/process concept is critical to rapid convergence of the process. Therefore, the most desirable manufacturing plan should be targeted from the start. Ideally, it should be predicated on the assumption that all design related obstacles can be overcome through clever design of the product and process as a system. Similarly, the design concept should also be "ideal" from a functional and user needs perspective assuming that all manufacturing difficulties can be overcome. By starting with the ideal, the design team has a clear idea of how to judge acceptability as the design concept and manufacturing plan are merged. The ideal design concept and manufacturing plan represent extremes that bracket the globally optimal solution and can, therefore, be used as a measure of "optimality" of the final integrated solution. By starting with the ideal or most desirable design concept and manufacturing plan, the team also has a constant appreciation for how information should best be distributed between product, manufacturing hardware, and computer software as the product/process design proceeds toward minimization of information content.

Though the DFM approach does not absolutely ensure optimal DFM solutions, if employed with diligence and intelligence, it will produce at least satisfactory results, and often excellent results. When redesign fails, the evaluation results together with the DFM evaluation help aid the search for an alternative geometric layout or design concept. Notably, if the DFM approach is employed early enough in the design, redesign seldom fails and most of the design team (or designer's) time and energy is spent creating a better, easier to manufacture and support design.

In most design situations, the DFM approach will be implemented concurrently with other design activities that are focused on functionality, reliability, and so forth. Alternatively, it can be utilized independently as part of a DFM design review or as a design improvement exercise or even as a means for trouble shooting a problem. The DFM approach is most definitely not a "trial and error" process. The path to the best design for manufacture is effectively pointed out by the results of previous evaluations, by qualitative physical reasoning, and by the principles of good design.

Examples that Illustrate the DFM Approach

The DFM approach is an attitude, a mindset, a philosophy of design that focuses on considering ease of manufacture and assembly early in the design process. It is best implemented using a team approach. As a minimum, however, the DFM approach requires nothing more than a philosophical commitment to DFM combined with use of the design-analyze-redesign process. The following examples illustrate this simple approach.

Example 5.1: Design of a Machined Part

In Chapter 4, the process plan for making the machined part shown in Fig. 4.7 (see Example 4.4) and reproduced here in Fig. 5.4 was discussed. Because the design is "frozen" (only engineering can make changes to the part design) upon release to manufacturing in Example 4.4, about all that can be done by manufacturing is to reduce processing time and manufacturing cost per part by experimenting with machining speeds and feeds. Typically, these manufacturing parameters are optimized by determining speed and feeds that provide acceptable surface finish and provide reasonable balance between cutting time and tool replacement time. In high volume production, reducing production time and manufacturing cost is important, so many manufacturing organizations expend considerable time and effort in such pursuits.

Suppose that the machined part of Fig. 5.4 has not been released to manufacturing and that the designer is free to change the design to simplify machining. The part shown is a dimensioned 2-D drawing that has been generated from a 3-D solid model and then annotated with the notes shown. This means that any design change can be implemented with a few keystrokes and mouse clicks. No engineering change notice is needed so making a design change is super easy. In other words, the change is being made early in the design when conceptual maneuvering space is wide with few irreversible design decisions to constrain it.

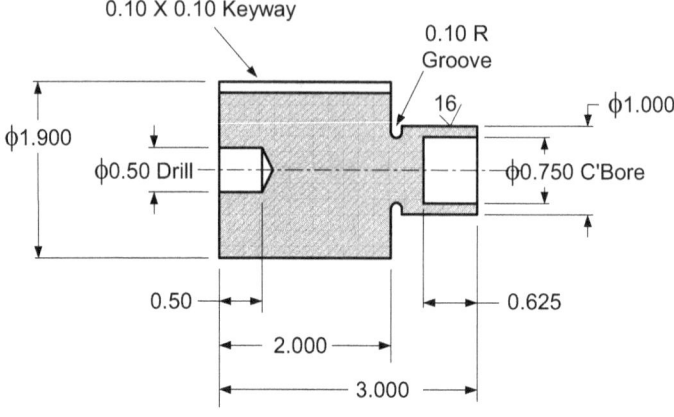

Figure 5.4 Production drawing of a part that is to be made in the firm's production machine shop (Chapter 4, Fig. 4.6).

Notes:
1. Material: CRS (cold-rolled steel).
2. All dimensions are in inches.
3. Diameters having "3-place" accuracy must be finished turned.
4. The 0.10-inch radius groove can be cut using a form cutter.

With this scenario, the part of Fig. 5.4 represents a first draft of the part design. The part is "ideal" from a functional point of view. Given this initial design, the designer seeks to simplify the part using the DFM approach. To do this, she calls upon the manufacturing engineer who is a member of the design team. The manufacturing engineer develops the "ideal" manufacturing plan shown in Fig. 5.5. On her own, it is doubtful that the designer could have come up with this plan. This illustrates the importance of the team approach and of having design information available when needed to facilitate high quality design decisions.

The tentative process plan requires 4 setups and 17 operations. Why so many? Does this part need to be so complicated? To answer these questions, the designer, perhaps in consultation with the manufacturing engineer, seeks to understand the reason for each separate setup and the large number of operations. This leads to the following insights:

1. Setups 1 and 2 are needed to obtain features F3, F4, F5 and F6. Because both F4 and F6 lie on the same centerline, a through hole would allow all four features to be machined from one direction. If a thru hole is permissible functionally, this would eliminate the need for the part flip and second setup (setup 2). Also, the higher cost of the CNC machine could be avoided.

The DFM Approach

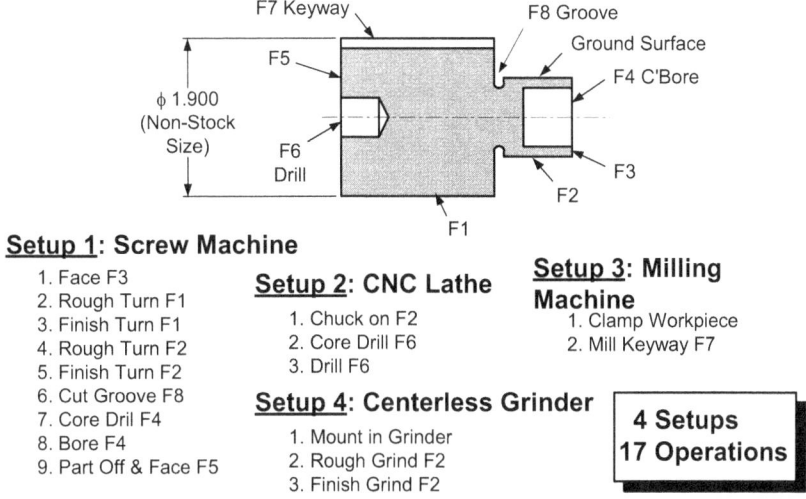

Figure 5.5 Draft process plan for the part of Fig. 5.4.

2. Given that the starting material is CRS (cold-rolled steel), the rough turn and finish turn operations performed in setup 1 to form feature F1 can be avoided by increasing the 1.900-inch diameter to a stock 2.0-inch bar diameter and specifying that 2.000-inch diameter precision ground bar stock be used as the starting material.

3. The purpose of the 0.10-inch radius groove (feature F8) is to allow "run-out" of the grinding wheel used to precision grind F2. Although this is considered good practice, it is not needed if the design of the part is carefully coordinated with the design of the mating part to ensure that no interference between parts occur.

With these insights, the part is redesigned as shown in Fig. 5.6. The process plan for the new design is shown in Fig. 5.7. Comparison shows that the redesign reduces the number of setups from 4 to 3 and the total number of operations from 17 to 11. This represents a dramatic reduction in design and manufacturing complexity, which translates into reduced total cost and total time. Table 5.1 presents a comparison of machining times and direct manufacturing costs for the before and after designs.

With respect to the DFM redesign, direct cost per part is reduced from $2.84 to $1.92 and cycle time from 6.20 minutes to 4.29 minutes. Subsequent speed and feed optimization produce an additional direct cost reduction of $0.16 and 0.34 minutes. These results clearly illustrate the benefits of the DFM approach. They also show that further process optimization is worthwhile but in no way compares with the DFM approach.

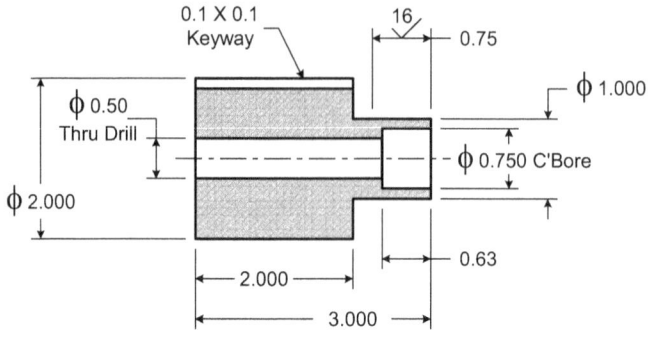

Notes:
1. Material: CRS (cold-rolled steel)
2. All dimensions are in inches.
3. Diameters having "3-place" accuracy must be finished turned.

Figure 5.6 Simplified part redesign.

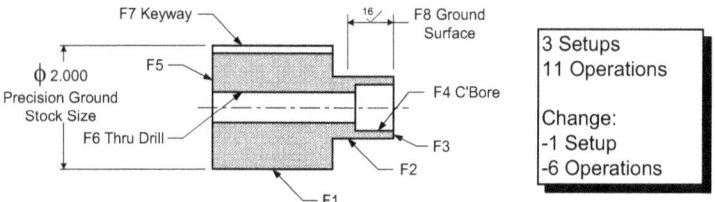

Setup 1: Screw Machine
1. Face F3
2. Rough Turn F2
3. Finish Turn F2
4. Core Drill F6
5. Bore F4
6. Part-Off & Face F5

Setup 2: Milling Machine
1. Locate & Clamp Workpiece
2. Mill Keyway F7

Setup 3: Centerless Grinder
1. Mount in Grinder
2. Finish Grind F8
3. Finish Grind F8

Figure 5.7 Process plan for the simplified part design of Fig. 5.6.

Table 5.1 Manufacturing Process Improvement.*

	Initial Design	DFM Redesign
Initial Speed and Feeds	$T_p = 6.20$ min $C_p = \$2.84$	$T_p = 4.29$, $\Delta T_p = -1.91$ min $C_p = \$1.92$, $\Delta C_p = -\$0.92$
Optimized Speed and Feeds	$T_p = 6.20$, $\Delta T_p = -0.49$ min $C_p = \$2.62$, $\Delta C_p = -\$0.22$	$T_p = 3.95$, $\Delta T_p = -0.34$ min $C_p = \$1.76$, $\Delta C_p = -\$0.16$

* T_p = machining cycle time per part, C_p = direct manufacturing cost per part

The DFM Approach

The results tabulated in Table 5.1 represent direct cost and time improvement. The reduction in indirect cost is considerably more significant. By eliminating one setup, all the time, effort, material waste, and quality risk associated with the setup has been eliminated. In addition, the CNC machine, a costly asset, is freed for more productive use and material handling is reduced since the part need only move between three machines instead of four. It is with these more intangible and harder to calculated indirect cost reductions that DFM proves its real worth.

Example 5.2: Design of a Simple Assembly

In a more conventional approach, the speed-stick assembly shown in Fig. 5.8 would be used by an assembly engineer to develop a process plan. One of the first steps in doing this is to decide on an appropriate assembly sequence and develop the assembly process accordingly. In doing this, the assembly engineer might imagine the operations that are to be performed, list them, and assign a "letter" designation to each possible operation so that alternative assembly sequences can be visualized. For the "speed-stick" assembly of Fig. 5.8, this list might be as follows:

      A – assemble receptacle and cap
      B – assemble receptacle and handle
      C – insert stick into receptacle
      D – place stick on cap
      E – place stick on handle

Although the assembly is composed of only four parts, in using this list, the assembly engineer finds that there are ten possible assembly sequences. For example, one assembly sequence might be A-C-B:

1. Fixture the cap (i.e., use the cap as the base part),
2. Assemble the receptacle and cap (operation "A"),
3. Insert stick into receptacle (operation "C"),
4. Assemble receptacle and handle (operation "B").

Alternatively, fixture the handle, assemble the receptacle, insert the stick, and assemble the cap (B-C-A). Or, fixture the receptacle, insert the stick, and simultaneously assemble the cap and the handle (C-A-B).

Some assembly sequences appear to be easier to perform than others. The A-C-B sequence, for example, might be much easier to tool and perform than, say, the A-E-B sequence. But the best sequence is not obvious and, in some cases, an assembly sequence that appears less promising may turn out

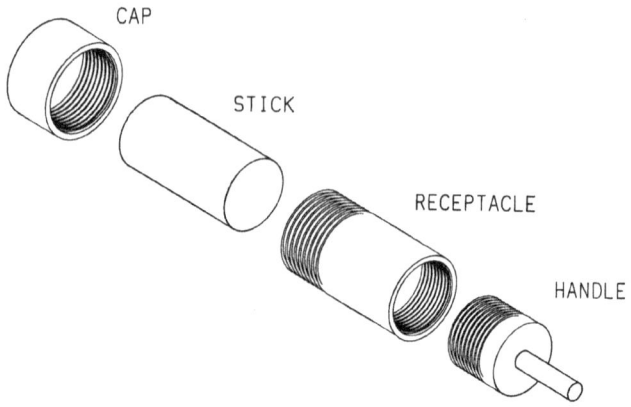

Figure 5.8 Speed-stick assembly.

to offer the lowest work content time, or the least number of workstations, or the best line balance. A detailed analysis of each sequence is required to know for sure. Without guidance from design engineering, the choice of assembly sequence is arbitrary, and the assembly engineer must select the assembly sequence that is deemed best. As an "at home exercise", the reader might find it informative to identify all ten speed-stick assembly sequences and then to decide which should be chosen.

The speed-stick example involves just four parts. Imagine the assembly engineering challenge of determining the best assembly sequence for products with higher part counts such as appliances and automobiles and the tremendous potential of the DFM approach becomes obvious. By considering the assembly sequence as part of the design process, ease of assembly can be greatly improved. Suppose that, in the speed stick example, the design team selects the A-C-B assembly sequence as the preferred assembly sequence. By knowing this sequence early in the design, "flats" or other suitable features can be added to the handle design to make it easy to clamp in a vice and hold vertically for "top-down" assembly, a taper can be provided on the appropriate end of the stick to facilitate insertion into the receptacle, and the cap can be designed for easy top-down threading onto the receptacle. Best of all, with the assembly design specified, assembly engineering is spared the onerous task of determining the best assembly sequence, identifying precedence constraints, and dividing the design into rational work elements. The result is design clarity and simplicity, which translates into significant direct and indirect cost savings.

Summary of Key Concepts

- ➤ The DFM approach seeks to ensure early consideration of "downstream" manufacturing and support needs by using concurrent engineering and the team approach.

- ➤ The design-analyze-redesign process used in conjunction with the design axioms and DFM guidelines, concurrent engineering, and the team approach underlie all aspects of the DFM approach.

- ➤ The DFM principles, together with use of structured methods and stretch objectives, help stimulate design creativity and innovation.

Chapter 6
Simplicity: The Holy Grail of DFM

Ask any engineer what the KISS principle is, and they will immediately tell you: "Keep It Simple, Stupid". The importance of simplicity in design is widely recognized. At the same time, achieving a simple design remains challenging. Why is this so? The problem lies in what is meant exactly by the term "simple". That is, how can the simplicity of a design be judged and how does the team know when it has achieved a truly simple design? In 1978, Nam Suh and his associates at MIT published a paper (Suh, 1978) in which they showed that the key to achieving a simple design is to understand how complexity can be measured and then minimized. With such a measure, the KISS principle transforms from ambiguous advice to a straightforward design directive.

The Information Axiom

Simple means not complex. *Information content* is a measure of complexity. Therefore, information content is a measure by which the simplicity of a design can be measured. More specifically, the lower the information content of a design, the simpler the design is. Therefore, to "keep it simple, stupid", the design team must strive to minimize the information content of the design because a simple design is one that has a minimum of information content. Originally called Axiom 2 (see Chapter 2), Nam Suh has restated it as the *information axiom* (Suh, 2001):

Minimize the information content of the design.

One way to think of information content is to imagine the set of instructions that would be required to delineate every aspect and detail of the product's manufacture, operation, use, maintenance, and disposal. Consider, for example, the number of setups and operations that are required to machine a given part. This is illustrated by the machined part example discussed in Chapters 4 and 5 (see Examples 4.4 and 5.1). The number of

Simplicity: The Holy Grail of DFM

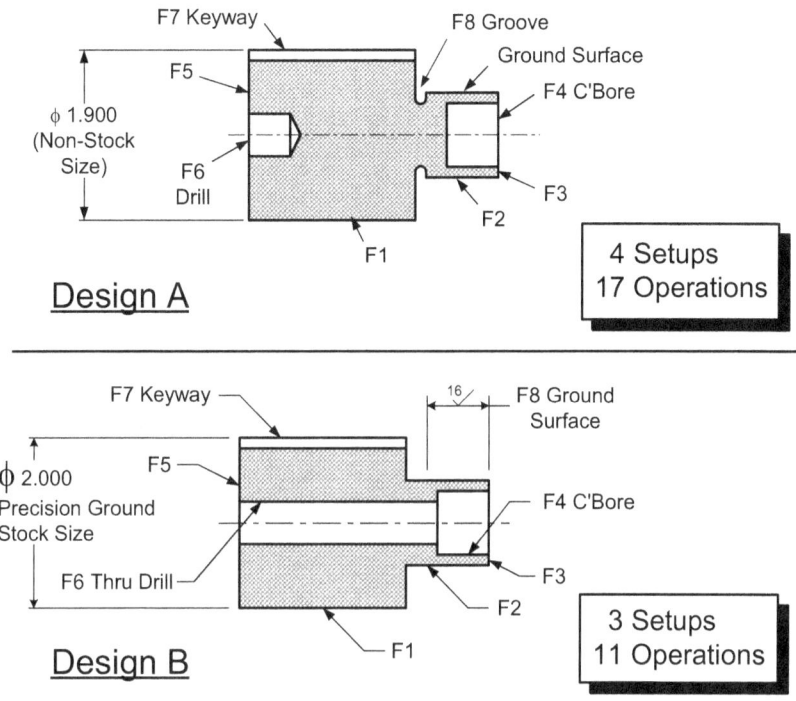

Figure 6.1. Setup and machining operation count for Designs A and B.

setups and operations required are shown in Fig. 6.1, where Design A is the initial design and Design B is the redesign created using the DFM approach. As shown, the redesign, "Design B", reduces the number of setups from 4 to 3, and the number of operations from 17 to 11. Therefore, Design B is "simpler" because fewer separate instructions are needed to make the part compared to Design A.

As this example illustrates, one of the really nice things about using "information content" as a measure of simplicity is that there is no uncertainty or ambiguity about what "simple" means. As a result, in many cases, reducing information content only requires imagination and creativity. Use a two-step approach: (1) begin by using your imagination to discern sources of information content; (2) for each source, creatively imagine ways for eliminating these sources and/or reducing their information content. Using this approach quickly leads to the sources and strategies listed in Table 6.1. Interestingly, the "simplify guidelines" listed in Chapters 9, 10, and 11, as well as many others published in producibility and value engineering literature and in company design brochures and

Table 6.1 Strategies that Minimize Information Content.

Source of Information Content	Design Strategy
Number of Parts	Design to minimize the part count. Design parts that remain to be simply shaped, easily made, and assembled.
Dimensions and Tolerances	Design to reduce the number of dimensions per part and per assembly. Avoid critical dimensions that depend on the assembly of multiple parts. Relax tolerances whenever possible.
Unique features, characteristics, functional surfaces, etc. contained in a component.	Design to reduce the number of each feature. Standardize and rationalize* alternatives. Develop a design with "standard features" approach.
Number of different tools and processes used in an assembly or in component manufacture.	Design to reduce the number of tools and processes. Standardize and rationalize* all tools and processes.
Number of separate activities; Number of steps per activity; Number of repetitions of each activity	Design to reduce the number of activities, steps, repetitions, etc. Standardize and rationalize.*
Randomness and variability.	Specify robust parameter values. Eliminate unconstrained components. Error-proof the design.

* To "standardize and rationalize" means to (1) reduce the number of standard options used in <u>existing</u> designs where possible, and (2) identify the lowest number of standard options for use in <u>future</u> designs.

pamphlets can also be quickly derived using this simple two-step approach. As Nam Suh and his colleagues discovered, the concept of minimizing information content underlies almost all design simplification strategies and approaches. For this and many other reasons, the information axiom is the "holy grail" of DFM. It makes the KISS principle straightforward, unambiguous, and ultra-easy to apply.

Simplicity: The Holy Grail of DFM

Sources of Information Content	Design A	Design B
• Number of new designed parts • Number of new vendors • Number of unique parts • Number of major new tools • Number of new production processes	12 3 2 2 0	5 2 3 4 1
Total	19	15

Figure 6.2 Example "complexity scorecard". Design B has the lower complexity score, and is therefore, the low information choice. Assuming all else equal, Design B would be the preferred choice.

To illustrate the power and convenience of the information axiom, consider the problem of selecting the best design concept from a group of alternatives. List the sources of information content contained in each alternative to create a "complexity score card" (Fig. 6.2). In addition to encouraging the team to consider all the sources of information content associated with each alternative, such a scorecard provides insight into trade-offs and intrinsic cost and is useful in helping to select the best concept.

Total Cost and Information Content

Simplicity is important because it is assumed that design simplicity correlates with low manufacturing cost. Is this, in fact, the case? By measuring information content, it is possible to test if there is a correlation, and hopefully, gain some important design insights along the way. As a starting point, it is necessary to calculate a numerical value for the information content of a given part or assembly design. One idea is to express information content (I) in terms of dimensions and tolerances (Stoll, 1999). With this approach, the information content of a design is calculated in terms of dimensions and tolerances as,

$$I = \sum_{i=1}^{n} log_2 \left(\frac{d_i}{t_i}\right) \quad (6.1)$$

where (d_i) is one dimension, (t_i) the tolerance on that dimension, and (n) the total number of dimensions required to specify the part. For assemblies, the dimensions and tolerances refer to the placement of each part in the assembly. Since information is usually measured in bits, the log is taken to the base 2.

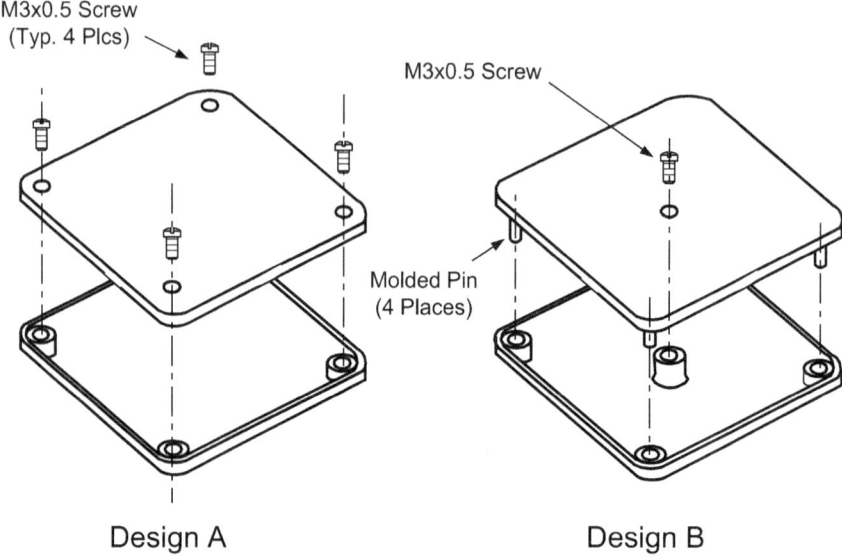

Figure 6.3 Alternative designs for securing a cover to an enclosure using threaded fasteners.

Using Eq. (6.1) and the simple assembly shown in Fig. 6.3, the correlation between information content and assembly cost can be tested. In Design A (Fig. 6.3), M3x0.5 screws are used to secure the cover. Tolerance on the screw diameter is ± 0.063 mm and the screws are tightened to a tolerance of $60°/360° = 1/6$th of a turn. The enclosure is 35 ± 0.5 mm in length on each side and the mounting holes are located on 30 ± 0.5 mm centers. The dimensional information content of the screw diameter is calculated as,

$$I_{Screw\ Dia.} = log_2\left(\frac{3}{0.063}\right) = \ln 47.619/\ln 2 = 5.57 \text{ bits}$$

where "log_2" is easily computed on a calculator using natural logs as shown. Performing similar computations for positioning the screw in the x and y-directions, tightening the screw $1/6^{th}$ turn, and adding this to the above calculation gives the dimensional information content for the installation of one screw as 22.56 bits. Similarly, the information content for placing the cover is calculated to be 12.26 bits. The total assembly dimensional information content of Design A is,

$$I_{Design\ A} = 4 \times I_{Screw} + I_{Cover} = 4 \times 22.56 + 12.26 = 102.50\ bits$$

Simplicity: The Holy Grail of DFM 97

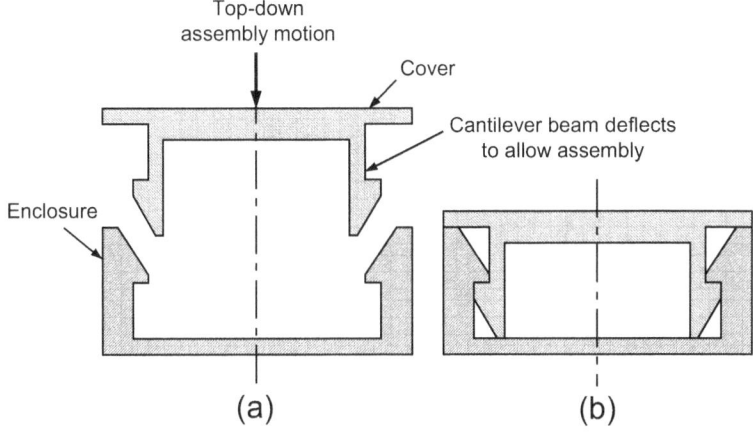

Figure 6.4 Snap-fit design (Design C). (a) Before; (b) After assembly.

In an alternative design, shown as Design B in Fig. 6.3, the cover is attached using a one-screw assembly with molded pins to locate the cover on the enclosure. The dimensional information content of this design, which has only one screw instead of 4 as in Design A, is calculated to be 33.3 bits. A third alternative, call it Design C, uses "snap-fits" to secure the cover to the enclosure. As exemplified in Fig. 6.4, "snap-fits" integrate the fastening and locating functions into the mating parts thus eliminating the need for separate fasteners. The dimensional information content for Design C, which uses no screws, is 12.3 bits, the information content for placing the cover.

The time to manually assemble these designs can be estimated using the Boothroyd-Dewhurst Design for Assembly Method (Boothroyd, Dewhurst, and Knight, 1994). In this method, the time required to handle and insert each part is estimated by analyzing the part geometry. These times are then summed to calculate the assembly time for the design. The resulting assembly times calculated for Designs A, B, and C are 53.33 seconds, 17.61 seconds, and 7.81 seconds, respectively.

A plot of total assembly time verses information content for Designs A, B, and C (see Fig. 6.5) shows that, as expected, assembly time decreases linearly with decreasing information content and that a strong correlation between assembly time and information content exists. Since assembly time is equivalent to assembly cost, this result confirms that information content is a proxy for cost. Therefore, it can be concluded that the information axiom is a reliable guide for making high quality early design decisions in the absence of detailed total cost information.

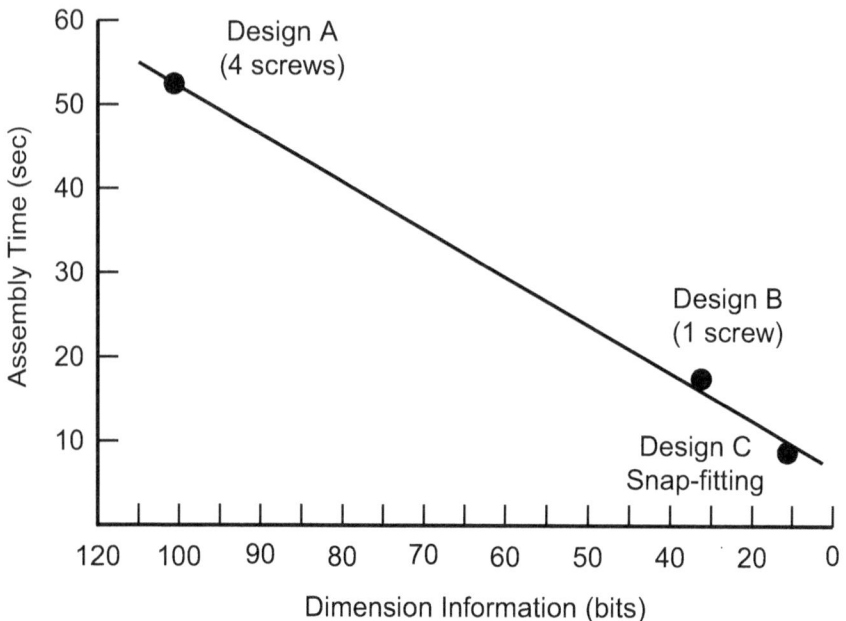

Figure 6.5 Plot of estimated assembly time verses decreasing information content for three alternative cover attachment designs.

Geometric layout and Information Content

The designs depicted in Figures 6.3 and 6.4 are examples of alternative geometric layouts. The lowest information content geometric layout that also satisfies the functional requirements of the physical concept should be chosen because it is the low-cost design. If the only requirement is that the cover be attached, the snap-fit design (Design C) would be the obvious choice. If the lead time and cost for the snap-fit tooling required is important, then Design B might be a better alternative, even though it is more informationally dense and therefore costly. If the cover must seal the enclosure to hold internal pressure or some other separating force, then Design A would likely be the only practical choice.

Of course, if holding pressure is a requirement, an alternative cover design that holds pressure while also minimizing information content, perhaps one that utilizes the pressure itself to help seal the cover, could be proposed (see Fig. 7.1, Chapter 7). There is no limit to the number of alternative geometric layouts that are imaginable. No matter what alternatives are under consideration, the idea of information content as a measure of simplicity makes it possible to identify low-cost choices.

Using the Information Axiom to Make Design Decisions

The amount of information contained in a part, subassembly, product, or production system, is determined by design decisions. Achieving simplicity "by design" in all things relating to the product and its method of manufacture helps guarantee the lowest possible total cost. This is a foundational principle of DFM. A design is "simplified" by making design choices that reduce its information content. For example, the separate screws used in Design A of Fig. 6.3 are sources of information content. By eliminating these screws in Design C, the dimensional information content of the design is reduced from 102.5 bits to 12.3 bits. And, this is just the dimensional information associated with assembly. What about all the information content required to purchase, receive, inspect, transport, store, handle, orient, insert, drive, and properly tighten the screws? By visualizing this additional information content, the design team can easily justify eliminating the four separate fasteners. This simple example clearly illustrates that it is not necessary to precisely calculate the information content when applying the information axiom, just mentally estimating it is enough.

The seemingly endless number of sources of information content in a simple screw can make it appear that estimating information content is no easier or more straightforward than calculating total cost. As we have seen, this is a fallacy, because unlike total cost, the exact amount of information content is not important. Rather, what is important is understanding how information content is affected by design decisions and using this understanding to make design choices. By simply understanding that information content is reduced when sources of information content such as separate fasteners are eliminated is all that is needed to have confidence in design decisions that eliminate fasteners.

Strategies for Reducing Information Content

Dimensions and tolerances are just one source of information content. In most design situations, there appears to be an almost limitless number of different sources. For products composed of discrete parts, for example, some of the more easily recognized sources of information content include:

- The number of separate activities, number of individual steps per activity, and the number of repetitions of each activity required to manufacture and/or assemble a product, subassembly, or component.

- The number of different tools and processes required in a product or component manufacture.
- The number of unique features, facets, characteristics, functional surfaces, and so forth present in a component.
- The number of interfaces and interactions between assembled components.
- The amount of part-to-part variability and build-to-build variability.
- The amount of randomness or hard-to-control variability associated with production processes, inspection processes, testing methods, material handling, order processing, shipping and warehousing, and all other activities associated with the production system.
- The amount of randomness or hard-to-control variability associated with product operation, servicing, maintenance, and disposal.

Given this long list of information sources, it is not hard to imagine an equation like Eq. (6.1) that would express the total amount of information contained in each different source. Suppose, for example, that there are (m) different sources of information content and that there are (n_j) individual components contributing to each source. Suppose further that the amount of information contained in each individual information source component is measured by the quantity (α_{ij}). Then, the total information content is,

$$I_{Total} = \sum_{j=1}^{m} \left(\sum_{i=1}^{n_j} log_2 \alpha_{ij} \right)_j \quad (6.2)$$

This equation is explicit regarding what must be done to reduce total information content: reduce the number of different information sources (m), reduce the number of components (n_j) that contribute to each source, and reduce the amount of information ($log_2 \alpha_{ij}$) contained in each component. This suggests three fundamental strategies for reducing information content of a design:

1. *Eliminate* sources of information content by reducing (m) and (n).
2. *Simplify* by reducing the information content ($log_2 \alpha_{ij}$) contained in each source component that remains.
3. *Standardize where possible* to further limit sources of information content (m) and (n), and the amount of information contained in each source ($log_2 \alpha_{ij}$).

The *eliminate* and *simplify* strategies are straightforward and relatively obvious. *Standardize where possible* is trickier because it decreases information content by reducing numerous options to a manageable few. If not carefully planned, however, this can inadvertently limit or constrain desirable options. The vagaries, dangers, and subtle traps inherent in standardization as well as the many benefits that it offers are explored in Chapter 8.

Summary of Key Concepts

- ➤ The information axiom is a truth for which no counter example has been found. This provides great confidence that decreasing information content will always decrease total cost and thereby help maximize total design value.
- ➤ The information axiom is the bedrock principle that underlies DFM. It provides a solid foundation upon which all design decisions, both large and small, can be based.
- ➤ The information axiom is best implemented using an eliminate, simplify, and standardize where possible strategy. The meaning of each term is explicit and unambiguous: *eliminate* means to remove sources of information content; *simplify* means to reduce the amount of information contained in an information source; and *standardize* means to limit the variety of sources.

Chapter 7
Undesirable Interactions: The Hidden Menace

Ever since the importance of design for manufacture was first recognized by American manufacturers, I have been walking different production lines to better understand how DFM can help. After doing this for a while, I found that I could actually "smell the information content (entropy)" on the line. This led to many insights regarding how designs could be improved to simplify their manufacture. But, in many cases, there was more going on than just excessive information content; there were quality and manufacturing problems that simply couldn't be written off as being caused by overly complex designs. Rather, something was inherently wrong with the basic geometric layout upon which the design was based. I became intrigued because I realized these flaws were a "hidden menace". Not only do they degrade total design value, and therefore profitability, they also derail the DFM approach by impeding its effectiveness. How can the information content of an inherently flawed design ever be truly minimized?

It was at about this time that I attended a seminar in which Nam Suh, a professor at MIT, discussed his ideas about "axiomatic" design. He began his lecture with a question "What was wrong with the idea of wanting to fly like a bird? After all, birds do indeed fly!" Professor Suh then went on to discuss the complexity of bird wings and all the functions that they perform over and above just flying. By trying to simply duplicate the geometrical shape and movement of the wing without understanding which functions of the wing are related to flying, early inventors were solving the wrong problem. The genius of the Wright brothers was to focus on identifying the functional requirements needed to fly. They identified three: (1) provide vertical lift, (2) provide directional control, and (3) provide thrust to propel the airplane forward. They then designed an airplane that did just these things. A wing was used to provide lift, a vertical stabilizer (the rudder) and "warpable" wing provided directional control, and the propeller provided needed thrust. What made the Wright brothers design successful is the fact that these three functional requirements were satisfied separately. In other words, the Wright bother's design "maintained the independence of functional requirements".

By studying many successful designs, such as the Wright brother's airplane, as well as less successful and failed designs, Suh and his colleagues formulated a fundamental axiom of design which they originally called Axiom 1 (see Chapter 2). Since then, Suh has restated the axiom as the "independence axiom", which is formally stated as follows (Suh, 2001):

Maintain the independence of functional requirements.

After learning about Suh's axiomatic approach to design, I quickly realized that the independence axiom explained the inherent but hard to explain design flaws that I had noticed when walking production lines. These flaws are the result of "undesirable interactions". Interactions are mutual or reciprocal actions or influences between different aspects of the design. They become "undesirable" when they interfere with functionality. Symptoms include unexplainable problems, degradation in product and/or manufacturing performance, poor reliability, unpredictability, loss of capacity, and increased uncertainty, inefficiency, and waste.

Excessive or difficult to explain and solve quality, manufacturing, and performance problems are "red flags" that signal loud and clear that an undesirable interaction may be present. Avoiding undesirable interactions requires constant diligence. Every decision must be questioned and tested for undesirable interactions. How could an undesirable interaction occur? What would be the symptom? "What-if" this? "What-if" that?

The obvious best way to avoid undesirable interactions is to use the independence axiom as a guide in designing both the product and the manufacturing plan. As discussed in Chapter 2, this requires that each functional requirement be fulfilled by a separate aspect, feature, or component of the design so that functional independence is maintained. Unfortunately, this is seldom the case in practice. More often than not, undesirable interactions are discovered as hard to explain problems that arise after the design has been launched into production and sold to customers.

Undesirable Interactions and the DFM Approach

When you have walked as many production lines and studied as many different product designs as I have, spotting undesirable interactions becomes a sixth sense. The same can probably not be said for the great majority of design and manufacturing engineers, and undoubtedly for managers as well. Although the independence axiom itself makes a lot of sense, for many, the concept of undesirable interactions that arise when the axiom is violated remains abstract, remote, and hard to pin down. It is for this reason that undesirable interactions are a "hidden menace". Once a problem that is caused by an undesirable interaction surfaces, untold amounts of time and money can be wasted on trying to fix it. One company that I have experience with had been fighting an undesirable interaction for more than two years. When I got involved, I spotted the problem almost immediately. They were astonished when I pointed it out because the cause was so obvious, if only they had known what to look for. Undesirable interactions are real threats that need to be taken seriously.

Almost all undesirable interactions are caused by a faulty geometric layout. Guarding against the possibility of undesirable interactions, therefore, must be a high priority in the early stages of design when the initial design concept is being developed and refined. As discussed in Chapter 2, one of the best ways to avoid undesirable interactions is to base design decisions on the independence axiom. This requires that the design team constantly question proposed geometric layouts looking for possible undesirable interactions. In many cases, undesirable interactions occur because of auxiliary FR's that are unique to the geometric layout. Therefore, be on the lookout for these FR's and possible undesirable interactions that could result. What would be the symptom? How could undesirable "couplings" occur? Doing this is admittedly difficult, but it becomes easier as insights and belief strengthen with experience. The payoff comes, of course, when a hard to explain performance problem, or quality problem, or manufacturability problem is avoided before it ever becomes a problem.

What should the team look for when they question a geometric layout? Each design is unique, so there is no general rule. Experience has shown that the most common cause of undesirable interactions is unwanted or unanticipated elastic deformation. Such deformation can be the result of forces developed either during product use or because of the manufacture and assembly process. Hard-to-control spring back, relative stiffness of redundant load paths, residual stress, tolerance stack-up, can all appear to be root causes, but in almost every case, it is the geometric layout itself that is ultimately at fault. Avoiding the "deformation" trap requires that force-flow be considered in every design decision. Follow the force-flow path and imagine how the component, structure, or assembly will deform under the action of the forces being applied and transmitted. Visualize the impact of this deformation on functionality. As illustrated in Fig. 7.1, when possible, use self-helping designs to combat undesirable deformation interactions.

Undesirable Interactions: The Hidden Menace

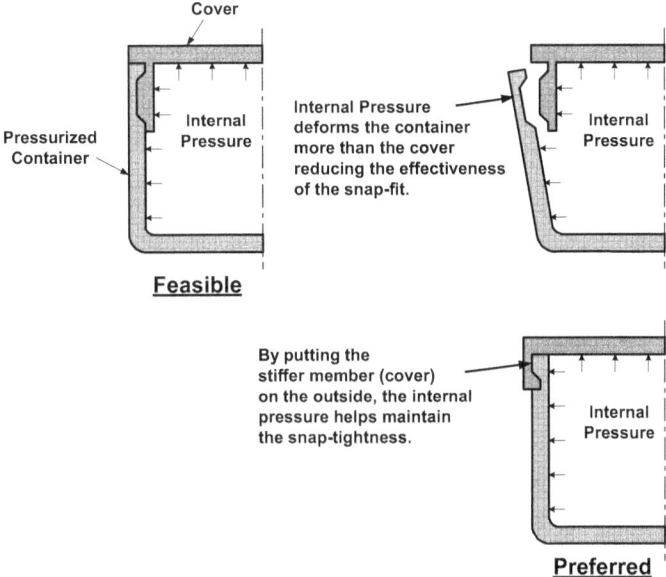

Figure 7.1 Example of "self-help" design.

Undesirable interactions can occur in many other ways in addition to unanticipated elastic deformation and distortion. Designs change for many reasons such as changing market needs, new product or manufacturing technology, changes in purchased component sourcing options, to mention just a few. Depending on the physical interfaces that exist between the product and the production system, these types of changes can trigger different unanticipated undesirable interactions, many of which wreak havoc on the manufacturing system in the form of part proliferation, tooling obsolescence, and a myriad of ripple effects that may occur. Interactions between part geometry, material, and manufacturing processes can be another major source of undesirable interactions. Concern for and awareness of undesirable interactions is a theme that pervades every aspect of design for manufacture.

Examples of Undesirable Interactions

Undesirable interactions manifest in many ways. The following examples typify the range of situations that can occur. They also illustrate how the independence axiom can be applied in creative ways to "decouple" interactions and, in some cases, provide easy "fixes".

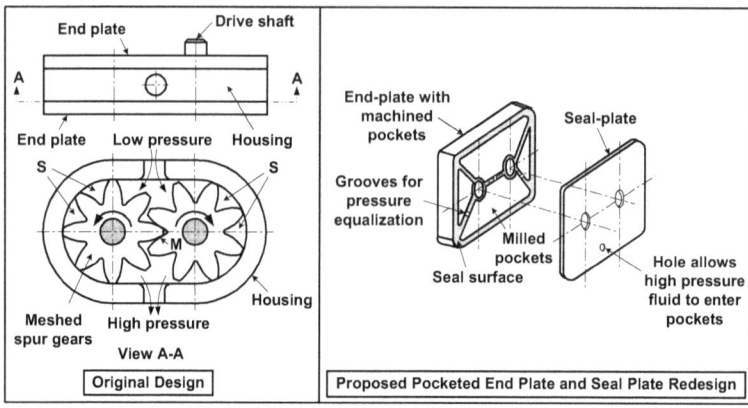

Figure 7.2 Leakage caused by endplate deformation is eliminated by the addition of a seal-plate that decouples the "contain fluid" and "seal pressure" functional requirements.

Example 7.1: Deformation Caused by Function Interdependence

Real materials deform under load. This deformation can interfere with functionality and result in undesirable interactions that degrade performance and capacity. When hard-to-explain problems with product performance occur and force-flow is present, suspect unwanted deformation. Visualization of the force-flow and consequent component deformation will usually reveal the cause of the problem. Consider, for example, the gear pump shown in Fig. 7.2. It works by trapping fluid in the spaces formed by the teeth of meshed spur gears and moving the fluid from the low-pressure intake side to the high-pressure discharge side. The spaces "S" between the gear teeth "contain" the fluid during this process. Since only a small fraction of the fluid returns through the gear tooth mesh at "M", high pressure develops when rotation of the gear's forces trapped fluid to flow against a discharge resistance.

The original design shown in Fig. 7.2 has failed to achieve specified discharge pressure due to fluid leakage between the endplates and the gear faces (gear side surfaces). "Beefing up" the endplates has not helped. When the team looks at this problem during a design-analyze-redesign cycle of the DFM approach, it realizes that an undesirable interaction exists between the "seal pressure" and "contain fluid" functions, both of which are, in part, provided by the endplates. This is because high fluid pressure acts to deform the endplates by causing them to "bow-out" thus allowing high pressure fluid to escape around the gear faces and flow back to the low-pressure intake side of the pump.

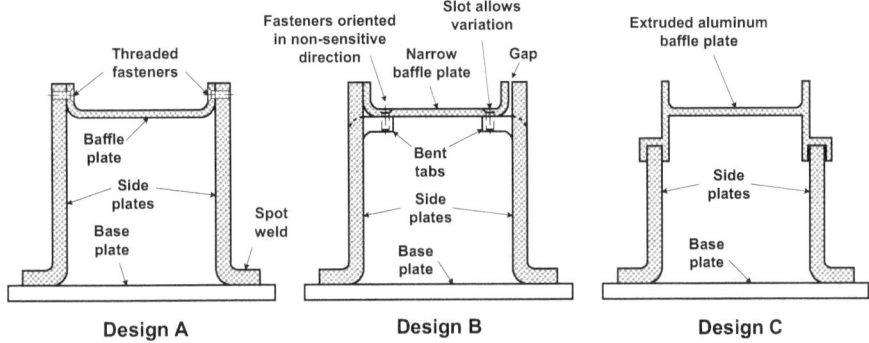

Figure 7.3 Cross-sectional views of three alternative geometric layouts.

With this insight, the team creates a redesign by adding "seal-plates" to decouple the "contain" and "seal" functions as shown by the proposed pocketed end plate and seal plate redesign in Fig. 7.2. In this proposed redesign, the endplates are still used to contain the fluid, but the "seal" function is provided by two additional "seal-plates". The seal-plates are "bowed" slightly so that when assembled, they act to provide initial sealing by pressing elastically like a spring against the gear faces. As pressure develops, a hole drilled in the high-pressure side of the seal-plates allows the pressurized fluid to flow via interconnecting grooves into pockets that are milled in the endplates. The pressurized fluid presses the seal-plates against the gear face with ever-increasing force as the pressure rises. This prevents the leakage path from ever forming and decouples the unavoidable deformation of the endplates from pump performance.

Example 7.2: Deformation Caused by Tolerance Stack-Up

Functional forces are not the only cause of unwanted deformation. For example, it can be caused by manufacturing realities such as tolerance stack-up. Figure 7.3 shows cross-sectional views of three alternative geometric layouts for a document transport mechanism. In Design A, hard-to-control dimensional variation results in an undesirable interaction. When the width of the baffle plate is too large, the side plates are spread apart. When the width is too small, the side plates are deformed inward as the screws are tightened. In both cases, deformation interferes with proper operation of the mechanism by causing binding of a drive roller shaft (not shown).

Without a doubt, it is possible to (1) fabricate a perfect baffle plate that has the exact width required, (2) form the side plates to the exact shape specified, and (3) precisely position and spot weld the side plates so that their spacing is exactly the same as the baffle plate width. In the case of

Design A, this is exactly what the model shop did. Because the model worked perfectly, the design was released to manufacturing and 40,000 drive motors were purchased at a bargain price. The problem was discovered when the product started to be returned from the field with burned out motors, poor speed control, and numerous other complaints regarding the transport drive system.

What the company learned the "hard way" is that the realities of mass production are far different from those of the model shop. The dimensions and shape of the baffle plate and side plates are hard to precisely control because of the inherent impreciseness of the sheet metal forming process. This is largely due to variability in material properties that affect elastic spring-back and other important behaviors. In addition, it is almost impossible to precisely locate, position, and maintain the required alignment of the side plates because localized heating produced by the spot-welding process induces distortion and other hard-to-control inaccuracies. The resulting hard-to-control variability leads to an uncertain product. In some cases, everything fits perfectly and the unit functions correctly throughout its service life. In other cases, baffle spacing and width don't match to varying degrees causing some units to fail completely while others exhibit marginal performance. No amount of rework, tightening of tolerances, or plant manager effort will fix the problem. The geometric layout of Design A is fatally flawed.

Since a redesign is required, the question now becomes one of selecting a geometric layout that avoids the problems of Design A and, at the same time, is also affordable. Both Designs B and C in Fig. 7.3 fill this bill, but each has different distributions of information content. In Design B, the baffle plate is mounted using exact constraint design (see Example 7.3 below) and is sized to always be narrower than the spacing between the side plates. In addition, the threaded fasteners are oriented so that they do not undesirably distort the side plates when tightened. In Design C, the multi-part baffle plate assembly is integrated into a single extruded aluminum part. Design B represents a "quick fix" because it is easy to implement and does not require more costly, long lead tooling. Design C reduces part count and isolates critical dimensions into a single part (the roller shaft bushings are mounted in the baffle plate instead of the side plates), but the extrusion tool required to make the baffle plate is more expensive. If time and money is limited or production quantity is low, then Design B may be the best choice. If, on the other hand, production quantity is high, Design C is preferred because information content is lower. One possibility might be to use Design B initially and move to Design C when production volume is sufficient.

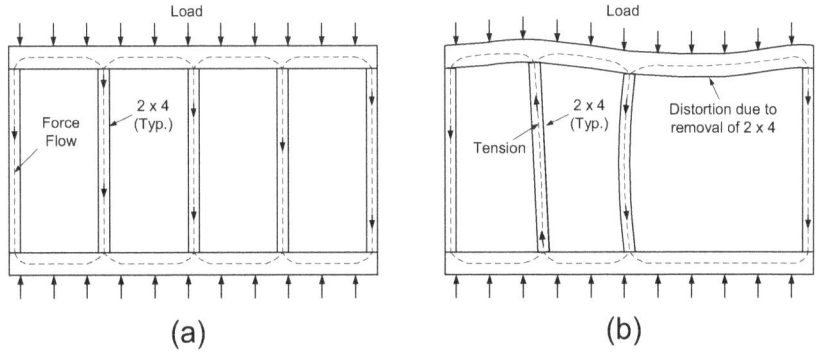

Figure 7.4 Force-flow in a load bearing wall. (a) Structure built according to design intent. (b) Distorted structure due to removal of one 2 x 4 load path (distortion greatly exaggerated).

Example 7.3: Deformation Caused by Over-Constraint

Over-constraint occurs when there are multiple force-flow paths that allow the force flow to distribute in indeterminate ways. This situation often arises in weldments, spot-welded structures, glued assemblies, and riveted assemblies, just to name a few. Force-flow is the path taken by a force as it passes through a component, structure, or assembly of parts. In general, force-flow can be determined by inspection using common sense. The force-flow in a load bearing wall (see Fig. 7.4a), for example, may pass through several vertical 2 x 4's. Because there is more than one load path, a load-bearing wall is a "redundant" structure. This means that one or more of the 2 x 4's could be theoretically removed without the wall collapsing, but this is difficult to do because the amount of load being carried by each 2 x 4 is hard to know due to the redundant load paths that are present.

Figure 7.4b illustrates the fundamental problem with over-constrained designs. As shown, one of the load paths has been deleted by the removal of one 2 x 4. To counteract the resulting imbalance in force-flow, the wall has distorted to keep the redistributed force-flows in static equilibrium. In the highly exaggerated illustration shown, one of the 2 x 4's has unexpectedly gone into tension (something it's not designed to do), another is on the verge of buckling because it is carrying a higher share of the load than intended and the distortion has caused it to become eccentrically loaded. This example clearly illustrates why it is unwise to mess with a load-bearing wall.

Figure 7.4 demonstrates the undesirable interactions that can arise due to over-constrained designs. When a geometric layout is chosen, implicit in the choice is the assumption that the design will be built as designed. But, in mass production, no part is perfectly made. Although the goal is to minimize variation, some variation will inevitably occur. In the case of the built-up wall of Fig. 7.4, each 2 x 4 is cut to length. No lengths are the same. This means that some of the 2 x 4's will fit tighter than others and therefore carry more load, at least initially, than the others. To compensate, the structure will distort until each 2 x 4 is carrying about the same load. The undesirable interaction is caused by this distortion. In some cases, such as the load bearing wall, no harm is done because the distortion required to redistribute load uniformly is hardly noticeable. In other design situations, such as optical design, the slightest distortion is unacceptable. Deformation caused by over constraint becomes a hidden menace when some product units meet specification while others don't, and no one can explain why.

Hard-to-control variation and the unpredictability that results is further complicated by the way force divides to flow through the multiple load paths. It turns out that load path stiffness is the key factor affecting force distribution; the stiffer, more rigid the load path, the greater the share of load it will carry. Consequences include the following:

- Parts fail for no reason.
- Problem seems temperature related.
- Tightening tolerances does not help.
- A thin component consistently buckles, warps, or distorts when assembled. Nothing seems to help.
- Reworking defective products sometimes does not help.

Over-constrained geometric layouts lack clarity, so they should be avoided if a more exactly constrained design solution is possible. One approach is to employ the principle of "exact constraint". An object is exactly constrained when just enough constraints are applied to unambiguously define its position in 3-dimensional space. The 3-2-1 fixturing principle discussed in Chapter 3 is an example. In this approach an assembled part is supported at three points in one plane, at two points in a second orthogonal plane, and at one point in a plane that is orthogonal to the first two planes. In Design B of Fig. 7.3, the baffle-plate is assembled by first placing it on three bent tabs that locate and support it in the horizontal plane. It is then moved to the left until it contacts two points on the vertical inside surface of the left side-plate. Finally, three screws are used to secure the baffle-plate to the bent tabs. These screws remove the final remaining freedom. Some redundancy remains, however, because which of the three screws is removing this freedom is not precisely known.

Undesirable Interactions: The Hidden Menace   111

Although exact constraint design is highly desirable when it is possible, in many design situations, rigidity and stability requirements render over-constraint a necessity. When this is the case, some "tricks of the trade" that are helpful for both identifying undesirable interactions caused by over-constraint and for eliminating them include:

1. **Understand the distribution of stiffness and strength:** Design so that stiffness (elasticity) and strength of each redundant load path is known. When possible, make the strength of each path approximately proportional to its stiffness.

2. **Understand how the design accommodates and absorbs unintended dimension variation:** Residual stress can be locked into a built-up structure during assembly due to deformation of components produced by dimensional variability, temperature gradients, and/or assembly forces. This stress can introduce unpredictable distortion. Prevent this distortion by making one force-flow path significantly stiffer than the others thus allowing tolerance stack-up, temperature induced shrinkage, and other dimensional changes to be absorbed by the less stiff, more flexible force-flow paths while still maintaining the shape and dimensions imposed by the stiffest force-flow path.

3. **Understand how tight tolerances affect the over-constrained design:** When there are redundant load carrying members, the more flexible members must undergo greater deformation to accommodate the smaller deformation of the more rigid members. Take advantage of this by making the stiffest, most rigid member also the most precise, dimensionally accurate member. In this way, the built-up structure will be dimensionally accurate because the flexible members will deform to accommodate the more precise rigid member. For example, in "skinned" structures such as those used in appliances, the skin, which is inherently flexible because of its thinness, will conform to the dimensions of the stiffer, underlying frame. Therefore, if the underlying more rigid frame is accurately made, the assembly will assume that accuracy.

Example 7.4: Assembly Problem Caused by Distortion

Figure 7.5 shows a simple assembly in which the base is clamped in an assembly fixture, internal parts are loaded "top-down" into the base and, as a last step, the cover is placed on the assembly to hold the whole package together. Only critical internal parts are shown for clarity. As shown in the figure, these include a rotating link, two pivot pins, and a tension spring. The

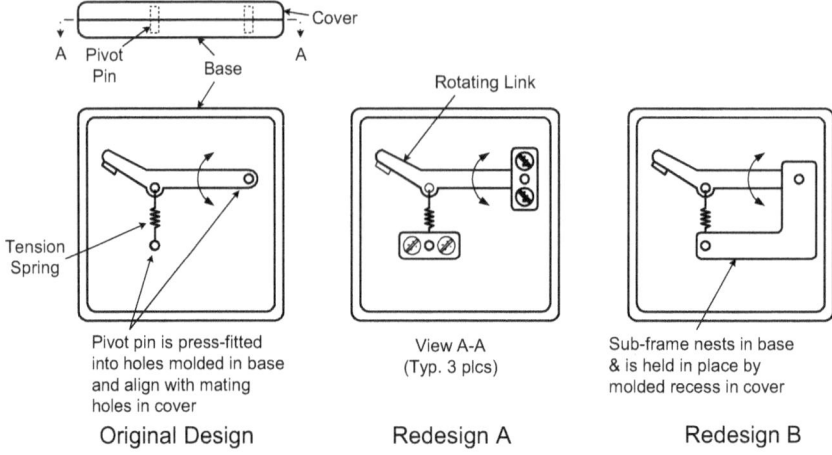

Figure 7.5 Spring-biased mechanism assembly (only critical parts shown).

pivot pins are retained in holes molded in the base using a light press fit. The link is free to rotate about one pivot pin and is spring-loaded by the tension spring that is attached to the second pivot pin. When the link is preloaded by the tension spring, the pivot pins tend to distort and the whole mechanism becomes an unstable "mouse-trap" that threatens to explode at any moment. The purpose of press-fitting the pivot pins in the base is to counteract the tendency for the pivot pins to distort. The decision to press-fit the pins is based on the "assumption" that the insertion force required was acceptable and was motivated by the designer's desire to keep piece-part cost low.

To work as designed, the pivot pins must remain sufficiently vertical so that they align with mating holes in the cover. Unfortunately, in practice, this isn't the case for a chain of unanticipated reasons. First, the press fit specification needed to be relaxed because the pivot pins are placed robotically, which requires a "light" press-fit to prevent excessive insertion force. In addition, the holes need to be chamfered to help guide pin insertion by a SCARA robot. To avoid schedule slippage, the light press-fit and addition of a chamfer design changes are quickly approved. With these changes in place, inherent hard-to-control variation causes the robotic assembly process to become unpredictable to the point where automated pin insertion is impossible. To correct the situation, the original design must be restored, and the robotic pin insertion replaced with a manual operation that is difficult and time consuming, even with special tooling. Making matters worse, each assembly must be individually tested to ensure proper operation.

In looking at this problem from the perspective of the independence axiom, we see that the problem is caused by an undesirable interaction between the "enclose" function provided by the base and cover and the "bias rotation" function provided by the tension spring. To maintain independence of functional requirements, these two functions need to be "decoupled". Two possible redesigns, redesign A and redesign B shown in Fig. 7.5, are proposed. Redesign A eliminates the undesirable interaction by moving the pivot pin assembly into the base alone. Alternatively, redesign B eliminates the interaction by independently satisfying the "bias rotation" functional requirement using a "sub-frame".

Both redesign alternatives eliminate the undesirable interaction. Which alternative should be selected? The answer can be found from the information axiom discussed in Chapter 6. Comparing information content of the two redesign alternatives, it is seen that redesign B requires the addition of one new part, the sub-frame, while redesign A involves several new parts. It could be argued that, even though it requires more parts, the parts are simple and low cost. However, this is offset by the increased assembly information content and quality risk. Although the sub-frame is more complex, it eliminates assembly work content (press-fit, critical alignment, difficult insertion) and the quality risk that accompanies it. If the functional requirements do not become coupled in undesirable ways, the geometric layout requiring the least number of parts is also the low information content design. Therefore, in the absence of more detailed information, redesign B is the preferred choice.

Summary of Key Concepts

> ➢ Undesirable interactions are surprises waiting to happen. In design and manufacturing, an unanticipated surprise, especially one that negatively impacts product performance or quality or ease of manufacture, can produce devasting, far-reaching, and long-lasting consequences.
>
> ➢ The independence axiom serves as a reliable guide for explaining and for decoupling undesirable interactions.
>
> ➢ Elastic distortion caused by faulty geometric layouts, and the tolerance stack-up and over-constraint that results, is the leading cause of undesirable interactions in assembled products.

Chapter 8
Standardization: DFM's Secret Weapon

Standardization is one of the most powerful and effective ways to simplify design and manufacturing. It works by limiting the number of available options and pre-defining the function, form, and fabrication descriptions of each available option. Standardization is effective at all levels of the manufacturing enterprise including components, assemblies, products, systems, materials, manufacturing processes, business processes, and organizational procedures. Because standardization limits options, however, it must be used carefully in ways that augment and support the business, product, and manufacturing strategy of the manufacturing enterprise. This means that, in general, the design team cannot, on its own, decide what and where to standardize. Effective standardization requires a strategic vision. Management at all levels must participate in planning the standardization strategy and, ultimately, there must be a companywide consensus that supports the plan.

Because of the serious cost and business consequence that might result, standardization is a potent DFM strategy that must be employed wisely. For example, it is often necessary to trade-off desirability of being a "full service" provider that is capable of meeting all needs against the desirability of reducing information content by limiting options. Charting the best path can be a daunting challenge, but it is not impossible provided a disciplined, thoughtful, and systematic approach to standardization is pursued. The purpose of this chapter is to provide insight and perspective on the trade-offs and opportunities that exist. The chapter begins by first systematically examining the benefits and costs of standardization. Following this, a systematic approach for identifying and assessing candidate standardization opportunities is suggested. Finally, several representative examples that illustrate how standardization can be used to eliminate information content are discussed.

Benefits and Costs of Standardization

Is standardization right for a given family of products or for a certain business or industry? When considering standardization, this must be the first question on everyone's mind. To answer this question, it is important to understand exactly what the benefits and costs will be. Table 8.1 lists the benefits and costs of some commonly used standardization schemes. These examples clearly demonstrate how standardization reduces information content as well as the costs that are imposed. They also exemplify the wide range of standardization opportunities available and schemes that are possible.

Benefits of Standardization

A well-conceived standardization strategy can yield many benefits. As illustrated by the examples of Table 8.1, benefits of standardization manifest in many different forms and tend to show up across the full spectrum of business and manufacturing activities in which the manufacturing enterprise is engaged. When evaluating a standardization opportunity, therefore, it is important that all aspects of the business be examined for potential benefits. Some of the most significant and effective of these include the following:

Increased Management Focus: By reducing complexity, standardization often allows management to focus more clearly and intently on the core business of the manufacturing enterprise.

Elimination of Setups: A "setup" includes all the non-value-added activity and material cost associated with changing a manufacturing process or operation from one product to another. Most setups involve adjustments, tool changes, material purges, scrap parts, and product rework, all of which generate direct labor, material, and equipment cost. Setups also generate indirect cost due to lost production, lost equipment utilization, and the scheduling effort required. Standardization avoids setups by allowing equipment to be dedicated to one product. In other cases, it can greatly simplify the changeover process resulting in reduced setup time.

Inventory Reduction: The inventory held by a firm can be divided into two categories: working stock and safety stock. *Working stock* is the amount of inventory held to meet the needs of a given lot size. *Safety stock* is the amount of additional inventory held to hedge against uncertainty. Standardization may significantly reduce both forms of inventory. When several product variants are replaced by one standard product, both the total number of parts and the variety of different parts that must be stocked are reduced. Experience has shown that reductions of up to 50% are possible.

Table 8.1 Benefits and Costs of some Standardization Schemes

Scheme	Benefit	Cost
Limit the number of product models (e.g., the Ford Model T)	• No setups • Less design time • Economies of scale • Management focus	• Product may be overdesigned and costly • No incentive for repeat buyers to upgrade
Limit the number of product options	• Fewer setups • Less design time • Less equipment cost • Reduced inventory • Management focus	• Product may be overdesigned • May loose customers if options are not available
Limit the number of feature options (e.g., hole sizes)	• Fewer setups • Fewer tools	• May have to redesign some products • May have to outsource non-standard options
Limit process and equipment options (e.g., only use spot-welding and buy all spot-welding equipment from one vendor)	• Reduce the range of training and special skills required • Reduce spare parts inventory • Reap benefits of being a preferred customer	• Lost capability may add cost and/or compromise functionality • Can't meet some needs
Use existing high-volume parts in new low volume products	• Save design time • Save tooling cost	• Part may be over designed • May require extra parts to adapt to new product
Limit purchased part options (e.g., reduce the number of standard screws used)	• Less design time • Tested and proven parts and processes • Established suppliers, favorable prices • Interchangeability between products	• Some components may be better than needed and therefore more expensive
Standardize manufacturing practices (e.g., standardize the die shut height of large stamping presses)	• Less setup time • Faster setup may lead to reduced inventory and improved JIT performance	• Cost required to change existing tools and equipment to conform with the standard

Simplified Material Resource Planning: Large product portfolios that include many product variants require costly material resource planning since each specific part used in each unique product must be stocked in the right quantity by correctly forecasting demand for each unique product. By increasing commonality among disparate product lines and models, standardization can greatly simplify the task of forecasting demand because the firm needs to only forecast demand for the whole product line, not individual products. Also, because the same parts are used in a variety of products, the danger of production delays due to part shortages is reduced.

Reduced Quality Risk: Standardization generally results in less of everything including less material handling, storage, production operations, together with less quality risk and ways things can go wrong.

Economies of Scale: Standardization frequently increases production quantity for standardized parts. This facilitates the use of more efficient production methods such as near net shape processes, simplified material handling and storage, and automated assembly.

Reduced Design and Development Time and Cost: Designers become more knowledgeable about the strengths, weakness, performance, and other salient characteristics of standardized parts. Design decisions are more straightforward, less testing is needed, and choices are reduced, all of which makes for faster and more effective design cycles.

Elimination of Cost Duplication: A product line composed of similar but not identical products can greatly increase costs associated with the design, production, and lifecycle support of each individual product. Each product variant may require its own set of tooling, its own design engineers, its own specialized maintenance and service procedures, its own testing and regulatory approval, and its own suppliers. Standardization will eliminate much of these cost duplications.

Beneficial Ripple Effects: The benefits of standardization ripple in many ways throughout the manufacturing enterprise. For example, many suppliers offer a quantity discount since larger orders reduce the supplier's cost by reducing setups and allowing the use of more efficient, less costly production methods. Therefore, if the standardization scheme involves a supplied part or component, a quantity discount may be an added benefit. In the same way, standardization benefits often cascade through the supply chain with the benefits of standardization being passed on from supplier to supplier. The whole supplier network should therefore be considered when analyzing potential benefits of a standardization scheme.

Costs of Standardization

As shown by Table 8.1, for most situations, standardization does not come free. Cost of standardization usually manifests in three ways:

1. **Overdesign:** something extra, such as unutilized material or added functionality, is "given away".
2. **Customer Needs:** the ability to satisfy certain customer wants or needs is curtailed or limited in some way.
3. **Capital Investment:** excessive up-front investment is required to implement the proposed standardization scheme.

Product models and variants proliferate to meet customer needs. Standardization can be used to reduce the information content of these situations. The challenge is to implement standardization without limiting the range of wants and needs that the product can satisfy. No matter how innovative a standardization scheme may be, it is often impossible to avoid "overdesign" altogether. A product is overdesigned when it is capable of better performance or offers more functionality than the application requires. The downside of overdesigned parts and products is that they may weigh and cost more than necessary. Extra weight is seldom desirable. In addition, the extra cost of overdesign increases as production quantity increases, making overdesign less and less acceptable for high volume products.

The consequences of overdesign obviously depend on the nature of the product. In the case of weight sensitive products such as transportation products (automobiles, planes), the extra weight penalty may be more than the market will tolerate. In other products, especially those where the overdesigned parts facilitate a desirable improvement in performance and functionality, an "upscale" version of the product may increase in popularity even though it costs more. In these cases, the cost penalty imposed by overdesign could eventually be offset by customer demand for the higher margin upscale product. This suggests at least two possible strategies for reducing the cost of overdesign:

1. Standardize in ways that don't result in overdesigned parts.
2. Standardize in ways that drive up demand for the least overdesigned products.

There are both external and internal costs of standardization. External cost arises because of market and competition factors. If standardization causes the firm to exit some market niches or causes the customer to prefer another competitor's product, revenue or market share may be lost. Unintended consequences such as these must be identified and understood. It is, therefore, extremely important that market niches and customer needs be carefully and thoroughly analyzed to understand the potential impact of standardization. For example, it may be logical to assume that the end customer will neither notice nor care that the same screw is used everywhere in the product. However, if this choice limits performance or prevents an important, but previously unrecognized customer need from being satisfied, then the unintended consequences could be high.

Investment cost is the primary internal cost of standardization. In almost all cases, there is design cost since the product must be adapted to the standardization scheme. In addition, tooling may be obsoleted, requiring that new tools and equipment be designed and procured. New plants and warehouses may be needed. Employee training or new worker skills may also be required. The savings produced by the standardization scheme must offset these costs. If the capital investment is very large, the payback period may extend over many years.

From Theory to Practice

Although standardization is highly desirable as a means for reducing information content, it does not make sense for every application. To be viable, the benefits offered by any standardization scheme must usually far outweigh the costs. In addition, even if the benefit/cost ratio is very favorable, the opportunity is still not viable unless the firm can afford the investment required. The challenge, therefore, is to identify and pursue standardization strategies and opportunities in ways that make solid business sense for the manufacturing enterprise.

Standardization opportunities exist wherever there is excessive information content in the form of excess options, manufacturing complexity, management complexity, inventory complexity, supply chain complexity, and so forth. All aspects and levels of the business such as raw materials, purchased components, in-process parts, and finished goods are potential sources. Business and manufacturing conditions that signal excessive information content include:

- Extensive proliferation of product models, designed parts, and purchased components.
- Lengthy testing and validation cycles for new product designs.
- Excessive inventory, long manufacturing lead times, inability to meet demand, scheduling difficulties, and other operational problems.
- Large numbers of different components purchased in small lot sizes.
- High overhead costs compared to competitors.

Depending on the nature of the product and industry, each company must approach standardization in its own way; there is "no one size fits all" approach. At the same time, the process of determining if and how to use standardization is essentially a four-step process.

1. **Identify Standardization Opportunities:** Understand all aspects of the business to identify promising opportunities. Form logical groupings of parts or products that have one or more characteristics or features in common. A logical grouping may consist of products that have a similar specification or functionality. Or it could consist of parts that are manufactured in a similar way or that have similar geometrical features or shapes. A *logical grouping* is, in general, any grouping of products, parts, features, or other characteristics that offers a potential standardization opportunity.

2. **Evaluate Qualitatively:** Develop a qualitative understanding of the costs and benefits associated with each candidate opportunity. Table 8.1, together with the information and independence axioms (see Chapters 6 and 7, respectively) are useful for doing this. It is also important to look for synergism as well as undesirable interactions that may occur if more than one of the candidate opportunities is implemented.

3. **Evaluate Quantitatively:** Develop a quantitative understanding of the benefits and costs associated with each candidate opportunity as necessary. Use this understanding to winnow, refine, and structure the candidate opportunities into a short list of the most promising and prioritize based on total design value improvement potential.

4. **Implement Promising Opportunities:** This pivotal last step is very difficult and challenging for many organizations. Identifying and evaluating reasonable standardization opportunities is relatively straightforward. Implementing them so that standardization works for the business and not against it is the challenge.

The goal is to identify and prioritize opportunities and to ultimately chart a wise course of action. To be successful, it is essential that many candidate opportunities be identified. By exploring the full spectrum of opportunities, it becomes possible to begin to see how standardization in one area might positively or negatively impact standardization in other areas. Also, by identifying many different opportunities, the possibility of identifying the most desirable opportunities is greatly increased. As in any idea generating activity, it is important to defer judgement until the big picture of benefits and costs has been clearly fleshed out.

Standardization: DFM's Secret Weapon 121

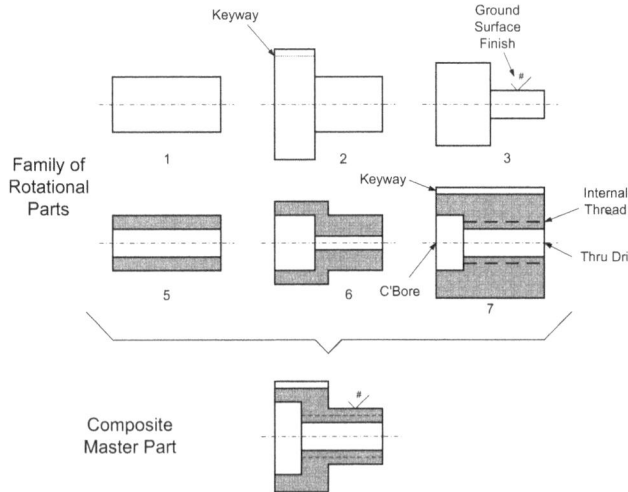

Figure 8.1 An example part family in which each part can be made by skipping one or more processing steps required to machine the master part.

Example Standardization Schemes

There are many ways to use the concept of standardization to eliminate information content. The approaches presented in the following examples are intended to be idea starters. The motivation for each is one of maximizing benefit while avoiding excessive cost and/or irreversible commitment.

Example 8.1: Use Previously Designed Parts

This scheme starts out by saving time and effort. Instead of designing a new part from scratch, new parts are created by modifying the CAD file of existing parts. Standardization of part families occurs gradually over time thereby minimizing downside consequences while also standardizing in the most logical and beneficial way possible. The scheme is implemented in five-steps:

1. Sort all existing parts into logical part families.

2. Create a master part for each family by consolidating all the features present in each version of the part into one composite design (see Fig. 8.1 for example).

3. Create a CAD library of master parts, one for each part family.

4. To design a new part, copy the most appropriate master part from the master part library and modify as required to create the new part.

5. By always using the same set of starting part designs, designers will naturally tend to simplify the design task by using the same features such as fillet radiuses and hole diameters in all new designs. For effective standardization, guidelines for preferred feature dimensions and characteristics should be established early. The guidelines should, in the general case, focus on the following goals:

   - Minimize the number of different part families.
   - Minimize the number of variations within each family.
   - Minimize the number of features used in each variation.

Applicability of this scheme is largely dependent on the nature of the products and parts involved. It is perhaps best suited for use in large companies that have many versions of the same general part. For example, there may be several to several hundred different "shaft" designs used in a large company such as an agricultural equipment manufacturer. Part families, such as a family of different shaft designs, are often obvious and can be quickly identified by inspection alone. In situations where many different products are involved, identifying part families can be more challenging. One approach that works well is to use "Group Technology (GT)" to capture part geometries in ways that facilitate identification of logical part families.

In addition to saving design time and facilitating part feature standardization, using previously designed parts as the starting point for designing new parts results in additional desirable benefits.

- **Standardization of Manufacturing Processes.** Because all parts in a part family share similar geometrical features, the process for manufacturing the part family can also be standardized. All parts in a family of machined parts, for example, can be produced by the same manufacturing cell or on the same CNC machining center using the same standardized tooling and setups. Switching from one part to another requires nothing more than computer generated changes to the CNC program.

- **Gradual evolution of "repeat parts".** *Repeat parts* are parts that are used in a variety of different applications. For example, a mounting plate or bracket used to mount a component in one product can be used for similar purposes in other products. Because it is often difficult to imagine how the same part can be used in multiple applications, repeat parts are usually discovered rather than intentionally designed. Using previously designed parts as a starting point for new designs facilitates this discovery process.

Example 8.2: Create a Catalog of Standard Common Parts

The parts that comprise the products of companies that produce and sell a variety of similar or related products and that are constantly introducing new and improved product versions generally divide into three categories: (1) unique parts, (2) common parts, and (3) purchased components. Handheld power tools exemplify these types of products. Functionally, an electric drill is totally different from an electric sander. From the perspective of parts, however, both products are similar in that they consist of housings that enclose common parts such as an electric motor, gears, shafts, bearings, and control switches, together with purchased components such as screws and cord sets.

In terms of parts, the power tool housing is a "unique" part because its shape and function is typically different for each power tool application. Therefore, in any new product development, a new housing must be designed to meet the needs of the new application. In general, *unique parts* are parts such as housings, brackets, chassis, and labels, that are configured differently, provide different functionality, and usually must be designed for each new application. *Common parts*, on the other hand, are parts such as shafts, bushings, bearings, gears, seals, and springs that have similar configurations and provide similar functions in all applications. In an internal combustion engine, the engine block, cylinder head, intake manifold, crankshaft, camshaft, and oil pan are unique parts. Conversely, the pistons, connecting rods, cylinder liners, wrist pins, tappets, valves, and valve springs are common parts. The third category of part, *purchased components*, include screws, washers, ball bearings, and all other components that are purchased from outside suppliers. Common parts differ from purchased components in that they are designed and manufactured internally by the company and are therefore exclusively used in the company's products.

The "catalog of standard common parts" standardization scheme is based on recognition that common parts, if standardized, can be used interchangeably in different applications. Such standardization produces increased flexibility and economies of scope and scale. Time to market is decreased because previous design and manufacturing experience with these parts reduces the need for extensive analysis and reliability testing. And, because the parts are already in production, tooling and production impacts are minimal. Standardizing common parts not only produces vast reductions in information content, it also frequently leads to increased opportunities to reduce cost further using automation and other productivity enhancement techniques.

Unique parts must usually be designed for each new product model or variant. By developing a catalog of standardized designs, the opposite becomes true for common parts. Rather than designing a new common part, the design engineer need only select from a list of proven designs. The challenge is to develop a catalog of designs that meet all conceivable needs and that can be used interchangeably in all future new products. To minimize information content, the catalog should seek to (1) minimize the number of different part categories (part families), (2) minimize the number of individual parts in each category, and (3) minimize the number of individual features that compose each part. Once developed, the catalog designs should be used exclusively in new product designs. Where deemed both cost effective and beneficial, it is also desirable that existing products be redesigned to accept the catalog designs. When performed in an orderly and systematic way, this will quickly free the company from making special, one-of-a-kind versions of the various common parts.

A general approach for developing a catalog of standardized common part designs is outlined as follows:

1. Survey all parts that have been designed by the company and are currently in production. Systematically sort these components into unique and common part categories.

2. Divide the common parts into logical part families. Develop a catalog of standard designs for each part family. In doing this, strive to minimize the number of variations at all levels to reduce information content.

3. Use the standardized common parts exclusively in all new design applications. Evaluate existing products, especially those that are not likely to be redesigned soon, on a case-by-case basis. Where cost effective and beneficial, make minor changes that will allow the use of the standardized components.

This approach facilitates a gradual transition to standardization that avoids the need for major redesign efforts or large capital investment. It is suitable in situations where the company's portfolio of products is successfully satisfying customer needs, the portfolio contains many existing product families and/or variants that are in production, and time and resources available for revolutionary changes are limited. As illustrated by the following examples, there are several variations to this approach.

Develop a "Master Part" for each Part Family. Standardizing common parts can undesirably restrict design freedom. To avoid this downside, rather than developing a series of standardized design variations

Standardization: DFM's Secret Weapon 125

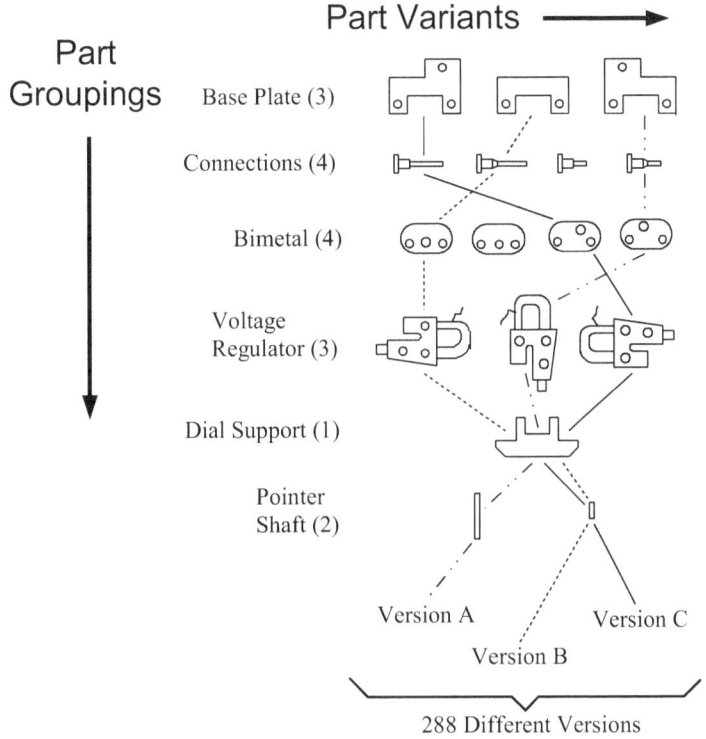

Figure 8.2 Automotive instrument gage building block design.

for each part family, develop a composite master part for each part family (see Fig. 8.1) and then develop a standardization strategy for each master part. For example, design so that any possible variation of the master part can be obtained by skipping some steps and features in the manufacturing process and using certain preferred hole diameters, fillet radii, and so forth.

Create a "Mix and Match Building Block" Design. When the standardization of common parts scheme is taken to its logical limit, the result is a "building block" design. In this approach, all the components comprising a family of products are standardized common parts. These parts are designed so that they can be mixed and matched in different combinations to produce all possible variants of a product family. Consider for example, the automotive building block design shown in Fig. 8.2 where mix and match standardized building block parts are combined to create up to 288 different automotive gages. When first introduced, about 500,000 gages, in 60 models, were produced every month with a cycle time of one second. The production line could be reset in just one second to produce a different model. This was done up to 200 times per shift. (Stoll, 1999). Often, companies that are already producing an existing product family are in the best position to create a successful building block design. This is because identifying the right building block groupings and variants typically requires extensive experience and insight. The goal is to minimize the number of part groupings and the number of variants in each grouping while maintaining the ability to meet all anticipated design requirements for several product generations into the future. A second goal is to design the building blocks to be combined in a way that allows flexible production of each product variant in lot sizes of one or more on one common production line.

Employ a "Revolutionary" Approach. Evolutionary approaches to standardization introduce standardization gradually on a case-by-case basis depending on benefits and costs involved. Revolutionary approaches typically involve a total reconceptualization and redesign of the product family. How this is done depends on the nature of the products and on the marketing and manufacturing strategies of the company. Consequently, there are many possible revolutionary approaches. In all cases, however, the product designs and manufacturing methods are likely to be entirely different (see Design Situation #1, Chapter 1). The upside of adopting a revolutionary approach is that the benefits of standardization can be fully leveraged. On the downside, revolutionary approaches often involve significant risk and require large capital investment. Revolutionary approaches are most appropriate in situations involving totally new products, new business startups, and in situations where some type of crisis such as new government regulations are forcing a total product redesign.

Design with Standard Features. If not implemented wisely, standardized common parts can significantly restrict the creative design freedom needed to satisfy customer, business, and manufacturing needs. The state-of-the-art of solid modeling and modern parametric CAD/CAM systems offers an alternative part standardization approach by allowing standardization at the "geometric feature" level instead of the "designed part" level. *Geometric features* are generic shapes or characteristics of a component's geometry. The number of geometric features used by the design engineer is small compared to the number of possible designs that can be created by using the geometric features in different combinations. By standardizing geometric features of common parts, many of the benefits of standardization can be reaped while avoiding the creative restrictions that can occur at higher levels of standardization. The design with features approach is still in its infancy and additional research and experience is needed before it becomes a widely used standardization scheme. For readers who find this idea intriguing, Shah, 1995 is a good place to start.

Example 8.3: Standardize Unique Parts

Unique parts, especially if they are large and/or have complicated geometry, can present considerable challenges in the design of some products. Lead times are usually long for these parts and the detail design of many other components and subsystems may depend on their detail design. For example, the detail design of housings, machine frames, and other complex parts need to be fully specified early so that the dimensions and configuration of the overall design can be determined. For these and many other reasons, it is often necessary to firm up the design of unique parts early in the program. If not carefully considered, however, these early design decisions can inadvertently constrain the design resulting in added cost and possible undesirable interactions. In the worst case, should the design of a unique part need to be changed, cost and time penalties can be high because tooling and many other parts are likely to be affected.

Platform Design. For most products, unique parts and common parts need to be treated differently. Unique parts tend to specialize or differentiate the product in one way or another. A hand-held electric power tool housing, for example, must be shaped to fit the application it is intended to serve. Consequently, unique parts must usually be specially designed for each new application. How then can unique parts be standardized in useful ways? One approach is to develop a "platform" product where a set of standardized, interchangeable "chunks" are shared across a range of the company's products. A *chunk* is any logical grouping of elements such as a subassembly, individual part, technology, or working principle that is amenable to standardization. Unlike a modular design such as a camera body and system of interchangeable lenses, platform designs are not usually recognizable by the end user. The platform approach is widely used in the automotive industry. For example, the Volkswagen MQB platform concept is currently used across at least 28 different vehicle models ranging from the Audi A3 to the Volkswagen Tiguan SUV. All MQB cars share the same standardized front axle, pedal box, and engine positioning "chunks", despite having different wheelbases, tracks, and engine positioning. In addition to greatly reducing vehicle design time, the MQB platform concept allows Volkswagen to assemble any of its MQB cars in any of its MQB ready factories thereby providing wide-ranging production flexibility. The concepts of "chunking" and coordinated product and process design are explored in greater depth in Chapters 14.

Design Rule Approach. Another highly effective approach to standardizing unique parts is to develop simple design rules that, when applied to each new design, simplify the design process and "decouple" the

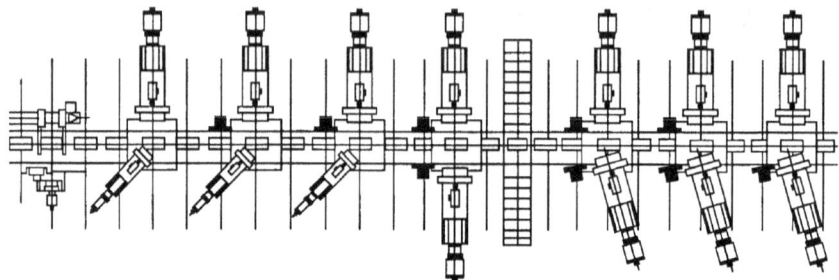

Figure 8.3 Graphical representation of a segment of a typical engine block finish machining transfer line. Such transfer lines are extremely complicated and long. If put on the walls of a large room, a complete drawing can extend completely around the room.

unique part design from the production process that is used to make it. This approach is often applicable to base parts, which are unique parts to which may other parts are added during the assembly process. To standardize the base part of a family of laser printer products, for example, it is only necessary to follow one rule: use the same base part envelop dimensions for all product models and variants. By following this rule for all models, ranging from the slow and inexpensive version for home use to the more costly, high speed office versions, one standard design can be developed. In addition, extra holes, "knock-outs" and other adjustable features can be included to meet all foreseeable needs and product variations. Because the base part envelop dimensions are standardized, the production line material handling system is independent of the product model being produced. This makes it possible to produce all product models interchangeably on the same line in lot sizes of one or more.

To illustrate application of the "design-rule" based approach to the design of an individual unique part, consider the design of internal combustion engine blocks for high volume automotive applications. Engine blocks are usually sand-cast using cast iron or light metal alloy and then finish machined on a large and costly transfer line. Figure 8.3 shows a graphical depiction of a short segment of such a line. The time required to design, build, install, and prove out a lengthy and complicated transfer line such as this can require years of effort. And, in many cases, when the engine is eventually phased out, the transfer line and associated tooling is usually obsoleted and never used again.

Standardization: DFM's Secret Weapon

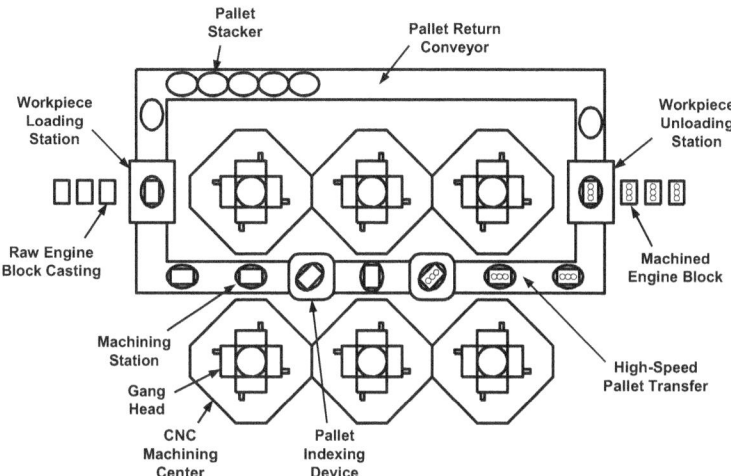

Figure 8.4 Flexible manufacturing system (FMS) for finish machining engine block castings. Each machining center machines in one orthogonal direction.

How can the "rule-based" approach be used to improve this situation? Figure 8.4 shows the layout of a flexible manufacturing system (FMS) for finish machining automotive engine blocks. Each machining station utilizes two fully programmable multi-axis CNC machining centers. The line can be quickly and easily changed over from one engine block to another simply by downloading the appropriate NC program. A requisite number of duplicate machining systems operate in parallel to satisfy production volume and cycle time requirements. Because each FMS is fully programmable, they never need to be obsoleted. New engine block designs are phased in by dedicating just a few of the FMS's to initial production. As volume demand for the new design increases, the old design is phased out by shifting more and more of the FMS's to production of the new engine design. The elegance of this solution lies not only in the fact that new engine block designs can be introduced quickly, but also in the fact that ramp-up to a new design can be managed easily and economically.

For this standardized finish machining solution to work, the engine block must be designed according to the following design rules:

1. The envelop dimensions of the engine block must be within prescribed limits.
2. All machined features must be obtainable using three orthogonal machining axes.

3. To reduce cycle time, holes and other features should be spaced to allow simultaneous machining operation.

Using these design rules to guide the design of all new engine blocks essentially standardizes the unique part design.

Example 8.4: Rationalize Purchased Parts

With respect to standardization, rationalization is a process of systematically identifying the fewest number of "preferred" options from the list of all possible options and then using the rationalized list exclusively in all new product designs. Benefits are three-fold: (1) reduced design time because choice is limited, and the company has extensive experience with all available choices; (2) increased purchasing power resulting from increased quantity; and (3) reduced storage, inventory, and material handling and storage complexity because of less purchased part variety.

Success of the purchased part rationalization approach depends on only using the rationalized options in new product designs. By doing this, all the information content associated with numerous options is gradually eliminated over time. In many cases, it may take several years for benefits to accrue, since considerable time is required for old products to be phased out and replaced with new designs that only use the rationalized purchased part options. To ensure success, long-term management vision is required and importantly, all levels of management must be committed to making it happen. In addition, it is essential that all stakeholders including design engineering, manufacturing, service, and purchasing be involved from the start. Without management commitment and broad, company-wide buy-in, purchased part rationalization efforts are almost always doomed to failure.

Because of the diverse interest and points of view involved, to be successful, purchased part rationalization depends greatly on following a well-planned systematic approach. The following five-step process is recommended as a starting point.

1. **Organize.** Decide on the scope of the effort. Form a team composed of representatives from all concerned organizations within the company. Obtain management buy-in and resource commitment.

2. **Research.** Systematically characterize all currently available options by seeking to understand (1) usage patterns, (2) the importance of differences between options, (3) why popular options are popular, and (4) why unpopular options are occasionally used.

3. **Rationalize.** Develop a rationalized list of the most popular options for use in future products. Analyze the rationalized list for unintended consequences and possible undesirable interactions. Look several product generations into the future to gage the effect on performance, functionality, and customer satisfaction. Anticipate manufacturing and technology change. Iterate as needed.

4. **Promote.** Present and sell the rationalized list broadly within the company. Obtain feedback and systematically analyze all criticisms and negative comments. Iteratively modify the list until all reasonable objections are addressed. This step is extremely important because each stakeholder organization within the firm must accept and commit without exception. It is essential that the entire organization understands that they may have to find creative new design, marketing, and manufacturing solutions, and in some cases, possibly sacrifice a marginal advantage or performance benefit.

5. **Implement.** Use the rationalized list of options in all new product designs. Management must be strongly committed in this step and must rigidly resist all requests for a deviation. It is important that the whole firm recognize that any deviation immediately defeats the purpose of the rationalization effort. Where possible, redesign existing products, especially those that are popular or are unlikely to be phased out any time soon, to only use the rationalized list of options.

The power of the part rationalization approach is illustrated by a case study presented by Stoll in Chapter 20 of the book *Product Design: Methods and Practices* (Stoll, 1999). In this example, a company is interested in using the part rationalization approach to reduce the number of different components it purchases. As part of organizing the effort, the company determined that it purchases 3617 different items divided into 24 categories ranging from electrical connectors to castings to plastic molding powders. Based on analysis of this extensive list of purchased parts and the opportunities for rationalization presented, the team chose to focus initially on cold-headed fasteners. This category of purchased part was chosen because it involved 359 fastener variants and a total purchased quantity of 62,832,000 fasteners per year. The distribution of fastener size, which was one of nine different attributes that characterized each fastener variant, before and after rationalization is shown in Fig. 7.5. As shown in the figure, the number of available fastener options was reduced from 359 to 44. It is hard to imagine the potential cost savings produced by this purchased part

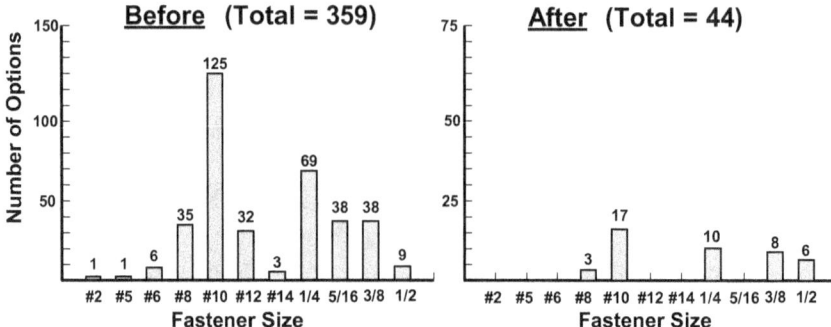

Figure 8.5 Fastener size distribution before and after purchased part rationalization. Total purchased quantity of all fasteners = 62,832,000 per year.

rationalization effort. However, this immense cost savings will only be realized if the company's management sticks with the rationalized list of options over the long-term. This need for long-term consistency highlights the importance of developing a company-wide standardization strategy that will sustain over the long-term and that will receive continued support and nurturing by all concerned.

Summary of Key Concepts

➢ Standardization is clearly a powerful tool in the DFM toolbox. Information content, and corresponding total cost, are dramatically reduced when standardization approaches are used to simplify part design and manufacture and to minimize proliferation of new part numbers within the firm.

➢ Benefits of standardization include:
  o Less of everything in the day-to-day operation of the business.
  o Less design time; faster design cycles.
  o Tested and proven parts and processes.
  o Established and favorable prices.
  o Interchangeability and portability.

➢ Potential downsides of standardization include overdesign, inability to satisfy some customer needs or market opportunities, and, in some cases, the need for sizable capital investment.

➢ To be truly effective, all standardization efforts must support and be in harmony with the firm's strategic vision.

Chapter 9
Part Count Reduction: The First Law of DFM

The primary goal of design for manufacture is to maximize total design value by minimizing the information content of the design. One of the best ways to do this is to focus on parts by using the "eliminate, simplify, and standardize where possible" strategies. These strategies are applied to parts by the three laws of DFM:

First Law: Eliminate as many separate parts as practicable.

Second Law: Design the parts that remain to be easy to assemble.

Third Law: Design the parts that remain to be easy to manufacture.

Unlike physical laws such as the laws of thermodynamics or mechanics, there is a natural sequence to the way the three laws of DFM are applied. First, eliminate parts, then design for assembly, and then, after the part geometry is fully determined, design the individual parts for ease of manufacture. At the same time, there is also an iterative give and take between the laws that allows the design to converge to the optimal minimum information content geometric layout. In this chapter, the focus is on the first law and the approaches for eliminating parts that are possible.

Elimination of separate parts is the single most effective way to eliminate information content from a manufacturing enterprise. Parts account for the great majority of cost, both direct and indirect. They are also the primary source of quality risk, unreliability, and customer dissatisfaction. A part that is eliminated costs nothing to design, make, assemble, move, handle, orient, store, purchase, clean, inspect, rework, or service. It never jams or interferes with automation. It never fails, malfunctions, or needs adjustment. No part number is needed. When products have been designed to minimize part count, they are inevitably and inherently easier to manufacture, assemble, service, and support.

If you step back from design and manufacturing and ask the question, "Where does the money go?" you will discover that part count drives all manufacturing cost. Divide the firm's total design and manufacturing budget by the number of individual components and parts that are manufactured or procured by the company and carried on its books and you will get an estimate of the average total cost per part, be it a screw or major casting. Now imagine what the design and manufacturing budget would be if the total part count were cut in half. It doesn't take a rocket scientist to see why reducing part count is the first law of DFM.

The Three Most Important Questions in DFM

In their pioneering work on design for assembly, Boothroyd and Dewhurst (Boothroyd, 1994) showed that a part can be evaluated regarding its potential for elimination by asking three critical questions:

1. Does the part move relative to other parts?
2. Must the part, for good reasons, be made of a different material?
3. Does the part need to be separate for assembly or service?

These questions are sometimes referred to as the "motion", "material", and "assembly/service" questions. They are the three most important questions in DFM because they are the key to eliminating the information content associated with excessive part count. If a part receives an answer of "Yes" to at least one of these questions, it probably cannot be eliminated unless a different design concept is used. For this reason, parts that receive at least one "Yes" are designated as *theoretical parts*. The *theoretical minimum part count* is the total number of parts in a design that receive a "Yes" to at least one critical question. Parts that receive the answer of "No" to all three questions, on the other hand, are *candidates for elimination* (CFE).

The three critical questions are not based on any rigid law of nature and are therefore subject to interpretation. To ensure that all redesign possibilities are explored, an answer of "Yes" should only be assigned for fundamental reasons. For example, the reciprocating piston of an internal combustion engine must move relative to the engine block for the engine to operate. Hence, the answer of "Yes" to the *motion* question is clear and unambiguous. A helical coil spring, on the other hand, is less easy to evaluate. Even though a spring must move relative to other parts to function, this relative motion usually involves elastic deformation, which means that some part of the spring can be stationary with respect to other parts. If only the stationary part is considered, then an answer of "no" to the *motion* question would seem appropriate.

It might be argued that the spring must be separate because it is made of a different material that needs to be fatigue resistant or have a certain yield strength. For the purposes of the critical questions, however, mechanical failure is an engineering concern and not a fundamental reason. Similarly, the steel cylinder liner in an aluminum engine block is needed for engineering reasons, not fundamental reasons. If an economically acceptable aluminum alloy with adequate wear characteristics were available, then the steel cylinder liner could be eliminated. Hence, the steel liner is a CFE and an answer of "Yes" to the *material* question for this part would close the door to innovative design solutions. An answer of "Yes" to an electrical conductor, on the other hand, is appropriate because the need to electrically isolate a conductor from its surroundings is an obvious and unambiguous fundamental reason for materials to be different.

The third question is usually the most ambiguous and difficult to answer. Purchased components such as ball bearings and light bulbs would receive a "Yes" to this question. So would the base part of an assembly since at least one theoretical part is needed to create an assembly. At the same time, should 25 light bulbs used in an automotive instrument cluster be considered as being theoretical parts when only one light source shining through a light pipe would work just as well? It is apparent that the critical questions of motion, material, and assembly/service must be considered carefully when deciding if a part is a candidate for elimination. My advice is to answer "No" when in doubt. This maximizes the number of candidates for elimination and acts to stimulate design innovation and creativity.

Although a theoretical minimum part count appears to be "ideal", this is not always the case. For example, part reduction should not exceed the point of diminishing return, where further part elimination increases the information content of the design by adding unnecessary cost and complexity. Red flags signaling that part count reduction has gone too far include parts that are too heavy, too complicated, or too unmanageable in other ways. In addition, part count reduction should not exceed the point where practical limitations imposed by performance requirements, development cost, and/or development time are exceeded. I have encountered many design situations where, for instance, using screws instead of snap-fits was preferable because of time to market constraints. Sometimes, companies must settle for less than "ideal" part counts because they can't afford the tooling required to eliminate some CFE's or because product sales volume is simply too low. In these situations, it is often wise to live with the more expensive design until sales volume and potential increases in cash flow justify additional investment. In any case, the part count should lie within the "efficient part count range" depicted in Fig. 9.1.

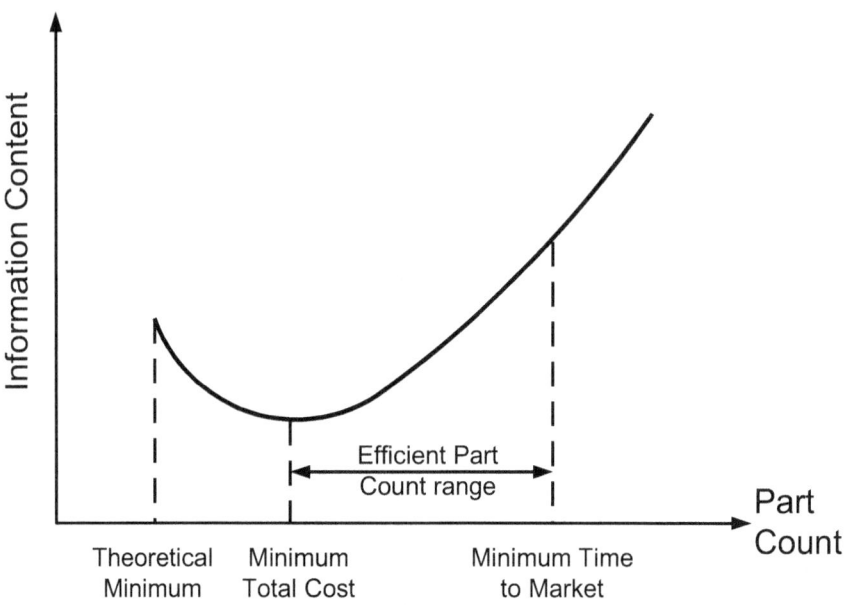

Figure 9.1 The "ideal" minimum part count may not correspond to the theoretical minimum number of parts if part geometries become too complex, if time to market is critical, or if resources are limited.

There are a variety of approaches for reducing part counts. In general, these can be divided into two categories: (1) part count reduction approaches applied during the product design process, and (2) longer-term approaches such as standardization that require company-wide strategic vision.

Design Process Related Approaches

Part count reduction is one of the main goals of the design-analyze-redesign process used in the DFM approach. A tentative or trial design concept is proposed and then analyzed for ways to reduce information content of the design using the information axiom and the eliminate, simplify, and standardize where possible strategies. A redesign is then created that embodies the information reduction ideas generated during the analysis step. This process continues iteratively until a minimum information content design that is acceptable to all stakeholders is achieved. One of the main goals of this process is to identify CFE's using the three critical questions and then finding creative and innovative approaches that can be used to eliminate them. In the following, we discuss several proven approaches for eliminating parts by design.

Part Count Reduction: The First Law of DFM

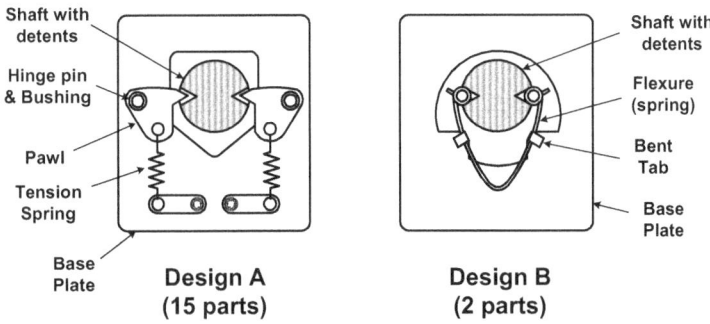

Figure 9.2 Two alternative design concepts for a shaft positioning device are shown. Of the two, Design B is preferable because it has only two parts and both are theoretical parts.

Reduce the Number of Theoretical Parts

The best way to eliminate parts is to identify a design concept that requires few parts (see Fig. 9.2). In some cases, this can be straight forward because the problem of design is relatively simple, and the minimum part count design concept is obvious or easily identified. In other situations that involve higher part counts and more complexity, however, identifying the minimum part count design concept usually requires a more concerted effort. The geometric layout improvement method presented in Chapter 12 provides an easy to use methodology for doing this.

Consolidate Parts into An Integral Design

Integral design, which involves combining two or more parts into one, is possible whenever two or more adjacent parts have been identified as candidates for elimination. In addition to eliminating the CFE's, integral design reduces quality risk since no fasteners or other joining processes such as welding or adhesive bonding are needed. Integral designs have fewer points of stress concentration, higher strength, and lower weight. An example of a single integral plastic part that replaces 12 separate parts is shown in Fig. 9.3. Plastics are a major key to integral design and are available for making springs, bearings, cam and gears, fasteners, hinges, and optical elements. Powder metallurgy (PM) is a good alternative if plastic parts do not have adequate strength, heat resistance, or cannot be made to the tolerances needed. Brazed, welded, or staked assemblies of stampings and/or machined parts can often be made as one-piece PM parts. Extrusions and precision castings are also excellent ways to eliminate subassemblies.

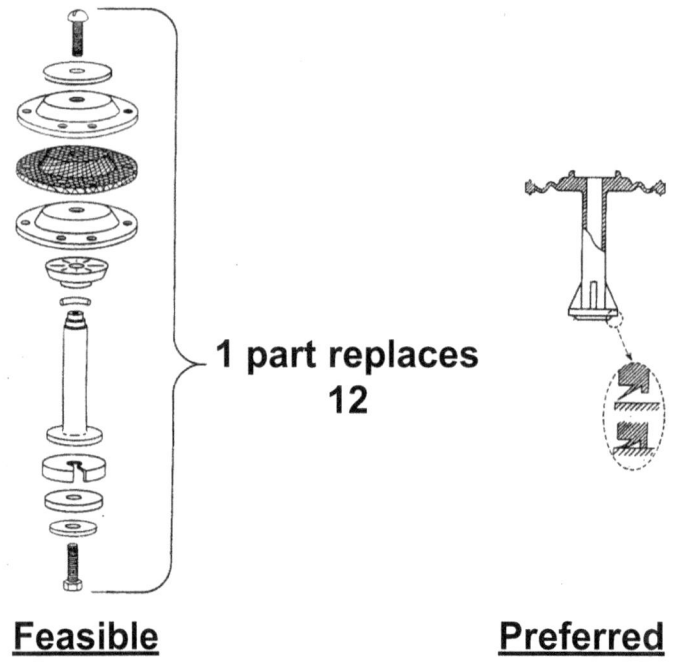

(Source: DuPont Hytrel Advertisement)

Figure 9.3 Illustrative example of integral design (Dupont Hytrel ad).

Although switching to a different manufacturing process may result in a costlier part, experience with part integration has shown time and again that a more expensive integral part turns out to be far more economical when assembly and indirect costs are included. Moreover, less assembly inevitably means less quality risk. The ease of manufacture and installation of modern automobile dash boards is an excellent case in point.

Design Hybrid Parts

When parts must be separate and integral design is not possible, *hybrid part design* offers an alternative. In this approach, assembly is incorporated into the piece-part production process. The hybrid part shown in Fig. 9.4 is illustrative of this approach. Other examples include a plastic molding in which threaded metal bushings have been inserted during the molding process and a shaft and gear that are joined in a die casting process. Often parts requiring plastics having different material properties can be combined by co-extrusion and other polymer processing processes.

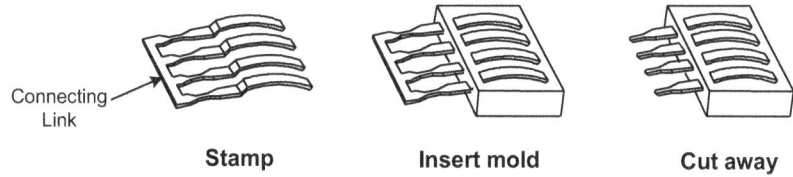

Figure 9.4 In this hybrid design, the need to handle and assemble numerous parts is avoided by postponing part differentiation. The multi-contact strip is fabricated as a single metal stamping. The stamping is then insert molded, after which the material connecting the contact strips is cut away.

Eliminate Separate Fasteners

Separate fasteners are always candidates for elimination since they will never receive an answer of "Yes" to any of the three critical questions. In addition, separate fasteners have many undesirable characteristics. In automated operations, fasteners can be difficult to feed reliably resulting in frequent jams and line shutdowns, and they require monitoring for presence and preload. They must be purchased, received, inspected, stored, moved to the point of use, and kept separate to ensure that the right fastener goes in the right place. If not properly installed or left lying loose in the assembly, they can present serious quality risk. Although the cost of most fasteners is relatively low, the information content associated with fasteners is very high. Eliminating fasteners eliminates indirect cost that can be six to ten times the cost of the fastener itself.

Integrating fastening functions into higher level parts using the principles of integral design is a highly effective way to reduce fastener count. Snap fittings are the most versatile and assembly friendly replacement. If separate fasteners must be used, the number of sizes and styles should be minimized. When possible, assemblies that utilize a single fastener are preferred (see Design B in Fig. 6.3, Chapter 6). In all cases, only standard, commercially available fasteners should be specified. Alternative permanent and non-permanent joining processes such as adhesive bonding, bent tabs, and press-fits can also be considered. Unfortunately, most of these alternatives come with their own set of issues that could increase information content rather than reduce it. Adhesive bonding is a prime example. It offers simple joint configurations, air-tight seals, high strength to weight ratios, and compatibility with dissimilar materials, but only if variables such as temperature, humidity, and time are properly controlled. In addition, special mixing equipment, dispensing techniques, and holding fixtures are often required.

Longer-Term Strategies for Eliminating Parts

Part elimination approaches such as standardization schemes and modular designs often require a company-wide top-down strategic vision. When properly implemented however, these strategies can produce enormous cost savings in the form of design and manufacturing simplification and economies of scope and scale (see Chapter 8). At the same time, because they frequently require a sustaining long-term "corporate vision", such strategic approaches can fail because they lack the long-term commitment and support required. These strategies therefore represent a significant design challenge requiring careful planning and execution.

Use Standard Off-the-Shelf Components

Use of standard, off-the-shelf components always reduces information content of the product and manufacturing system. Unless the firm is very vertically integrated, a stock item is almost always less expensive than a custom-made item. Standard components require little or no lead time and are more reliable because characteristics and weaknesses are well-understood. They can be ordered in any quantity at any time. They are usually easier to repair, and replacements are easier to find. Use of standardized supplied parts also makes it possible for the supplier to become a part of the design team. Standard components can also simplify field repair and servicing. For example, common standard parts can often be purchased locally. In emergency situations, it may also be possible to scavenge common standard parts from other products to effect quick repairs.

The many advantages of standardization can be further leveraged using the concept of rationalization discussed in Chapter 8. At one company, the computer printout of standard washers available for selection by the designer was 14 pages in length and contained 448-part numbers (Bradyhouse, 1987). Although the various washers were arranged in size order, it took the designer considerable time to select the washer because of variations in material, finish and thickness. A new, rationalized list was developed that contained only 7 washers with bore sizes ranging from 3 to 16 mm. Choice was confined to only one material (steel), one finish, and one thickness. This rationalized list has been loaded into the standard library in the company's computer system so designers can call up the file and quickly copy and paste the chosen washer into their design. The washers have also been sized to complement the company's new rationalized bolt sizes.

In addition to significantly reducing part count, purchased part rationalization is a substantial time and cost saver:

- Parts are quickly selected from a compressed list that is well organized.
- The selected part has already been tested, saving a great deal of time. This is particularly true for parts requiring life tests, such as switches and ball bearings.
- Time and effort do not need to be expended locating a suitable supplier and negotiating a favorable price.
- Standard parts generally are volume parts allowing significant economies of scope and scale. For example, suppliers may choose to locate a manufacturing facility or supply depot near the company purchasing its product in high volume.

Design Parts to be Multi-Functional

Two important corollaries to the independence axiom (see Chapters 2 and 7) that apply generally to design and manufacture and have important implications for significant reduction in information content are stated as follows:

- Integrate functional requirements into a single physical part or solution if they can be independently satisfied in the proposed solution.
- Decouple or separate parts or aspects of a solution if functional requirements are coupled or become coupled in the design of products and processes.

These corollaries suggest that functions should be combined whenever possible in order to eliminate parts and simplify the design. For example, a part can be designed to act as a spring, an electrical conductor, and a structural member provided that these functionalities aren't coupled or don't become coupled in undesirable ways. Less obvious combinations of function might involve adding guiding, aligning, and/or self-fixturing features to a part to aid in assembly of the part. Or, perhaps a reflective surface or recognizable feature might be added to facilitate vision inspection or aid a robotic insertion operation. These latter examples illustrate inclusion of functions that are only needed during manufacture. Such function combinations are frequently the result of the DFM approach and its focus on including downstream manufacturing needs into the early conceptual stages of design.

Design Parts for Multi-Use

Many parts can be design for multiuse. For example, the same mounting bracket can be designed to mount a variety of components. The same gear can be used for different applications in different products. A spacer can also serve as an axle, lever, or standoff. Multiuse parts reduce cost by reducing the number of different parts or part variations that need to be manufactured. They also produce economies of scale because of increased production quantities of fewer parts and economies of scope because the same part is being used in a variety of applications and products.

A systematic process for creating multiuse parts is suggested in Example 8.1 of Chapter 8. This approach has been used by one company to review 14 commodity product families that it manufactures (Bradyhouse, 1987). Of the 1919 parts contained in this group of products, 235 were identified as common parts for use in future designs. Another 250 parts, such as nameplates and external housings, were identified as unique parts. The remaining 1434 parts, or about 75% of the total number of parts examined, were classified as proliferative parts that will be excluded from future designs. Projecting this against the company's current parts database of 50,000 numbers, it is anticipated that this approach will eventually lead to 6,000 standardized common parts, 6,500 unique parts, and 37,500 candidates for elimination.

Minimum Part Count Assessment

The "minimum part count assessment" is an objective assessment of how well a design succeeds with respect to the part elimination DFM goal. It is also illustrative of structured methods that allow the design team to objectively evaluate the manufacturability of a design quantitatively and then use the results to improve the design. Such methods provide motivation and lead to insights that stimulate creativity. The basis for the assessment is the part count efficiency, which is defined as:

$$Part\ Count\ Efficiency = \frac{Theoretical\ Minimum\ Number\ of\ Parts}{Actual\ Number\ of\ Parts} \quad (9.1)$$

A systematic, step-by-step procedure for calculating the part count efficiency is presented in Chapter 12. In a nutshell, this procedure consists of determining the actual number of parts by first taking the product apart (or imagining how that would be done) and then reassembling it in the reverse order. The theoretical minimum number of parts is determined by asking the three critical questions as each part is added to the build.

To illustrate how the part count efficiency might be used, suppose an assembly consisting of seven separate parts is analyzed and it is determined that three of these parts are theoretical and must be separate, while the other four are CFE's. The part count efficiency would be 3/7, or 43%. Suppose further that the insights gained in performing the analysis stimulates design creativity that leads to a proposed redesign of the product that requires the three theoretical parts plus one CFE for a total of 4 parts. The part count efficiency would then be ¾ or 75%. This improved score provides an objective indication of the information content reduction potential that can be expected if the redesign is put into production.

The "minimum part count assessment" is a good starting point for improving the geometric layout and for formulating strategies for reducing information content. Some examples include the following:

1. In many designs, the physical concept or technological approach used determines the minimum number of parts. Consider modifying the physical design concept or changing the technological approach to reduce the theoretical minimum number of parts even further.

2. Often, a part becomes a theoretical part by virtue of the geometric layout that has been implemented. Seek innovative geometric layouts that convert theoretical parts into CFE's.

3. Reconsider alternative design concepts. Determine the theoretical minimum number of parts for each concept. Consider adopting favorable aspects of alternative concepts.

4. Look for ways to reduce the number of "extra" parts needed to provide differentiation between currently produced product/model variations.

5. Check all parts for function and modify the design to eliminate redundancies.

6. Assess all product models and variations currently in production. Look for common CFE's and theoretical parts. Consider developing a common design strategy that can be used across all products and their variants to minimize the total corporate part count. Seek ways to standardize large parts as well as unique parts that are similar across different product models or variants.

Cautionary Notes

Part count reduction is a key DFM goal. It is important, however, to always keep in mind the pitfalls and dangers that surround part count reduction. Although some of the following cautionary notes have already been sounded, it never hurts to reiterate them again.

- When carried too far or inappropriately applied, part reduction can increase rather than reduce information content of the design. Be on a constant lookout for unanticipated consequences and second order effects. Most importantly, guard against undesirable interactions.

- Standardization can be a particularly effective strategy. However, be sure to fully understand the cost/benefit trade-offs when considering standardization as a strategy for eliminating parts (see Chapter 8).

- Some parts have higher value than others. For example, a component that has been deliberately designed as a "standard common part" and that is used interchangeably in several products has inherently more value than a unique part that must be specially designed and tooled for one application. Elimination of high value parts should be carefully considered, especially if doing so necessitates the design of one or more low value parts.

- Once quantitative methodologies like the "minimum part count assessment" become accepted, management may be tempted to set design goals (e.g., part count efficiency of 80%). Such goals can be beneficial by providing additional basis for judging alternative design concepts. But it is important to remember that the minimum part count assessment is a design tool and not an end in itself. What good is it if the team hits the target goal by designing a product that is hard to manufacture and increases rather than decreases information content.

Summary of Key Concepts

➢ Reducing part count through integral design, elimination of fasteners, and so forth is perhaps the single most effective means for minimizing information content of the design.

➢ The optimal part count balances the theoretical minimum number of parts against complexity, development cost, and time to market.

➢ Total corporate part count can be reduced by developing visionary design strategies involving standardization, rationalization, platform design, commonality, to mention a few.

Chapter 10
Design for Assembly: The Second Law of DFM

When I first became involved with design for manufacture (DFM), I had the opportunity to walk an automotive assembly line that was set-up to assemble two different vehicles, each designed by a different automobile manufacturer. Call these Company A and Company B and their respective vehicle designs Vehicle A and Vehicle B. When the line was configured to assemble Vehicle A, one build after the next moved along the assembly line, with one car coming off the assembly line every time the line cycled. When the line was configured for Vehicle B, on the other hand, each vehicle build was followed by three empty spaces on the line, so that one car came off the line once every forth cycle. In other words, Vehicle B required four times the assembly effort of that required by Vehicle A. Why was Vehicle A so much easier to assemble? Both had an engine, four wheels, a dashboard, front and back seats, four doors, a trunk, and so forth. The answer is plain and simple: Design for Assembly (DFA). I have never forgotten this lesson.

Most "showstopper" manufacturing problems occur during assembly and are usually discovered during product launch. When this happens, it can have devastating cost and timing impacts. For this reason alone, it is important that assembly be given close attention during design. But, as the vehicle assembly example illustrates, this isn't the most important reason. When assembly is considered early in the design, cost, quality, and productivity all benefit. Design for assembly (DFA) is key no matter the production quantity.

Design for assembly begins with part count reduction. A part count of one requires no assembly. Unfortunately, this is seldom an option because of the many inherent reasons for assembly. Relative motion, different materials, standard off-the-shelf components, service and wear part replacement all require separate parts. Assembly simply cannot be avoided in most designs.

So, how should a product be designed to minimize assembly information content? In this chapter, we seek to answer this question by (1) identifying and understanding the sources of assembly information content, (2) using this understanding to develop a set of design for assembly guidelines, and (3) illustrate each guideline with examples.

Design for Assembly Guidelines

Analysis of the assembly process shows that, in general, adding a separate component to a "build" will involve some or all the following basic operations. Each of these operations represents a source of information content because they require time, tools, equipment, and involve a variety of non-value add material handling and checking activities.

- *Handling*: the process of grasping, transporting, and orienting a component to ready it for insertion into the build.
- *Insertion*: The process of adding a component to the work fixture or partially built assembly.
- *Securing*: the process of securing a component to the partially built assembly either immediately upon insertion, or later as part of a separate operation.
- *Adjustment*: the process of using judgement or other decision-making process to establish the correct relationship between assembled components.
- *Separate Operation*: mechanical and non-mechanical joining processes involving parts already in place but not immediately secured following insertion. Examples include bending, upsetting, screw tightening, resistance welding, soldering, adhesive bonding, to name a few. Also included are special assembly operations such adding liquids, preloading springs, and so forth.
- *Checking*: the process of determining that handling, insertion, securing, and adjustment has been properly performed.

Assuming that the part count is greater than one, the following design for assembly guidelines are formulated to minimize assembly information content of the design.

1. Design parts for easy handling.
2. Design parts for easy insertion.
3. Design for component securing.
4. Eliminate or simplify adjustments.
5. Avoid separate operations.
6. Error proof the design.

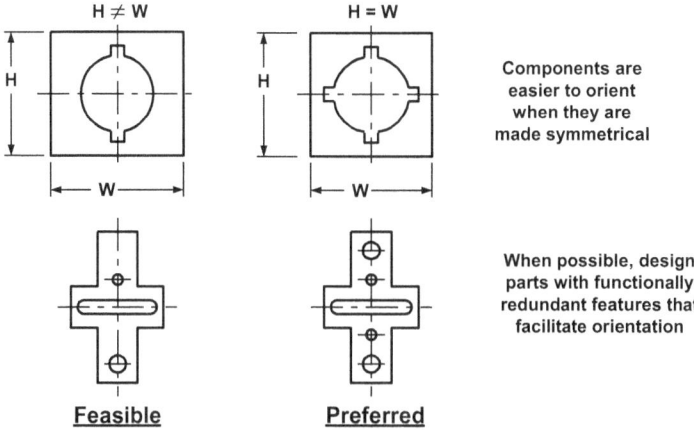

Figure 10.1 Symmetrical parts are easier to orient.

Design Parts for Easy Handling

Handling involves positioning (x, y, z) and orientating (α, β, γ) parts. Achieving the correct position and orientation costs money. Therefore, parts should be designed to make position and orientation easy to achieve. The number of orientations drives equipment expense, quality risk, feed rates, and cycle times. To minimize handling cost, parts should be symmetrical (see Fig. 10.1). If this is not possible, then asymmetry should be accentuated and made obvious. As a last resort, a clear identifying mark should be provided. Orientation can also be assisted by designing in features that help guide and locate parts in the proper position. When possible, design to prevent nesting or tangling in parts bins or vibratory bowl feeders. Similarly, avoid parts that are sharp, slippery, flexible, or fragile. Prevent shingling (the tendency of parts to climb up on each other) in bowl feeders by providing thicker contact edges or vertical or highly angled surfaces. Robotic part handling can be facilitated by providing a large, flat smooth top surface for vacuum pickup, an inner hole for spearing, or a cylindrical surface or other feature with enough length for gripper pickup.

Because parts usually come off the production line properly oriented, this orientation should be preserved by using magazines, tube feeders, or part strips. Palletized trays and kitting are methods for supplying properly oriented parts to the assembly line. Features should be designed into the product to facilitate packaging. It is also important to consider material flows within the production facility including product flow, workspace flow, part supply flow, hardware flow, trash and scrap flow, bulk material flow, container flow, and fixture flow.

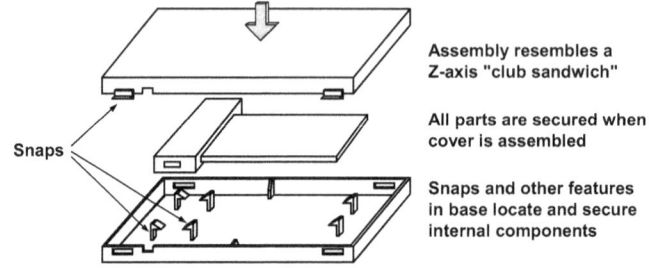

(a) The ideal assembly is a layered Z-axis stack.

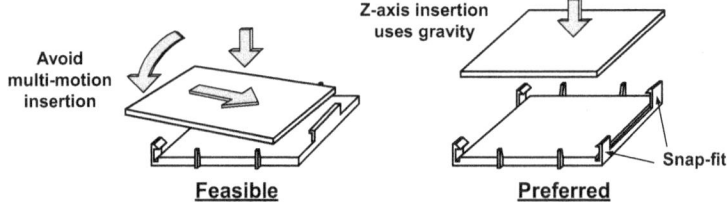

(b) Design so that insertion motion is as simple as possible.

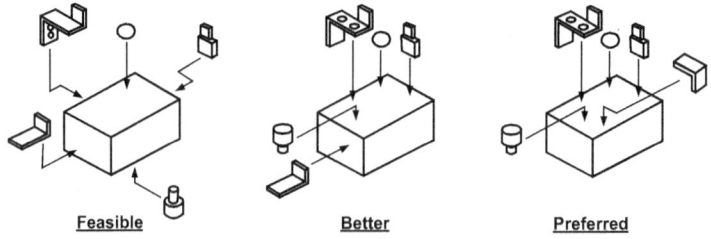

(c) Reduce processing surfaces.

Figure 10.2 Assembly is facilitated by simplifying assembly motion and reducing the number of surfaces to which parts are added. For multi-surface assembly, design so assembly is completed on each surface before moving to the next.

Design Parts for Easy Insertion

Parts will be easy to insert into the build when: (1) the motion required is minimized, (2) guiding features that align and guide the insertion process are provided, and (3) access to the build is unrestricted.

Minimize Assembly Directions and Motions: Parts should be assembled from one direction (see Fig. 10.2). Extra directions waste time and motion, and often require more processing stations. This, in turn

Design for Assembly: The Second Law of DFM

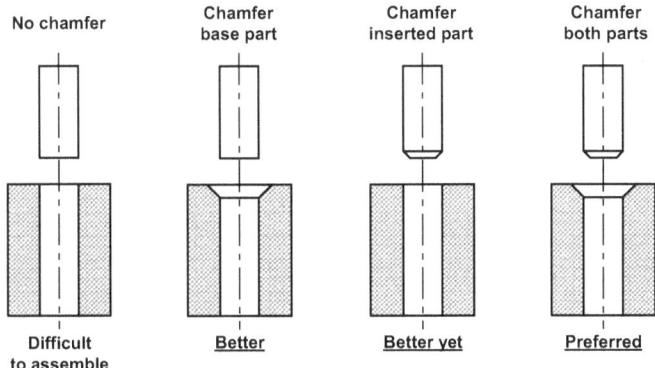

Figure 10.3 Provide generous chamfers, tapers, and radii.

increases cost, unreliability, and wear and tear on equipment due to added weight and inertia. The best possible assembly is when all parts are added in a top-down fashion to create a Z-axis stack. Multi-motion insertion should be avoided. Ideally, the product should resemble a Z-axis "club sandwich", with all parts positively located as they are added to the build.

Provide Guiding Features: Because parts are not always identical and perfectly made, misalignment and tolerance stack-up can produce excessive insertion force. Major factors affecting rigid part mating include part geometry (accuracy, consistency), stiffness of assembly tool, stiffness of jigs and fixtures holding the parts, and friction between parts. To compensate, compliance that allows minor alignment adjustments to occur during the part insertion process should be built into both the product and the production process. The simplest solution consists of a combination of (1) consistent quality parts; (2) designed-in part features such as generous radii, tapers, leads, and chamfers that align, guide, and help center the part (see Fig. 10.3); and (3) a base component that is rigid and accurately located. If automation is involved, it should have built-in selective compliance that facilitates insertion alignment and minimization of insertion forces. SCARA (selective compliance assembly robot arm) robots are widely used for this purpose.

Ensure Adequate Access, Clearance, and Unrestricted Vision: Open access, ample clearance, and unrestricted vision are essential for easy assembly and minimization of quality risk. This guideline seems so obvious and yet you would be amazed at the number of products that I've analyzed that fail this test. In some cases, the need for adequate access is simply overlooked or missed during design review. In other cases, such as large complex machinery, fulfilling this guideline is often a difficult challenge for which there is no easy solution. It is not unusual for access problems to be

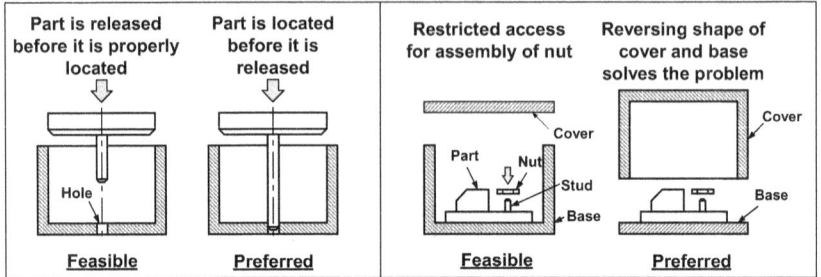

Figure 10.4 Improving access early in the design.

discovered late in the project when fixes are difficult or impossible to implement. In these cases, the problem and resulting cost and quality risk must be "lived with" for the life of the product. To avoid such disasters, strive to anticipate possible difficulties by visualizing the assembly process early in the design when changes are still relatively easy to make. Never fail to ensure adequate clearance for standard assembly tooling and readily available tools. Finally, make this a question that always gets asked during design reviews. Figure 10.4 illustrates some typical access problems that are easy to fix if discovered when design changes are easily made.

Design for Component Securing

Component securing is the process of physically attaching components to the partially built-up assembly using permanent or non-permanent joining processes. Securing may occur as part of the insertion process (e.g., installation of a threaded fastener) or it may be performed as a separate operation (e.g., adhesive bonding). A component is designed for easy securing when it is located and retained upon insertion and requires no screwing or plastic deformation or hard-to-control operation as part of the securing operation. Snap-fits, circlips, spire nuts, and so forth are examples of components that are easy to secure.

Many securing operations increase assembly information content and should be avoided when possible. For example, if a component needs to be held in place while it is being secured or if it can move or shift position during the securing operation, then extra information content is required making the operation more difficult and the outcome less certain. Similarly, if a joining process requires excessive force, time, effort, or special skills, it will, of necessity, involve extra information content. Adhesive joining processes require time and effort to properly distribute the adhesive and cure the bond. Critical welds must be inspected. Riveting involves plastic deformation and therefore requires special tools, equipment, and handling.

Design for Assembly: The Second Law of DFM

The integrity of many joining processes is also suspect. For example, the number of spot welds are often increased beyond what is needed to guard against a hard to detect defective welds.

Design guidelines for minimizing information content associated with component securing are best stated as prioritized optimal suggestions:

1. When possible, design so that as components are added to the build, they are (a) correctly located and oriented by fixture nests or other features on mating parts, (b) they do not need to be secured immediately, and (c) they do not need to be held in place by an external means. Ideally, the final part should secure all components using a snap-fit (see Fig. 10.2a).
2. If the component is secured immediately upon assembly, avoid screwing operations or plastic deformation of part features. Snap-fits are preferred when possible.
3. If plastic deformation is required, then metal bending or torsion is preferred to riveting or similar "upsetting" operations.
4. Use dog-point screws to avoid cross-threading. Avoid separate threaded fasteners and washers when possible.

In all cases, joining process uncertainty or excessive time should be avoided. If this is not possible, then the firm should invest in the process to develop it as a core manufacturing expertise.

Eliminate or Simplify Adjustments

Manual and automated mechanical adjustments are expensive and a continual source of assembly, reliability, test, and service problems. In addition, equipment that goes out of adjustment is a serious cause for customer dissatisfaction. Avoiding adjustments reduces assembly cost, enables automation, and improves service while also reducing service cost. The need for adjustment can be avoided in a variety of ways. One example is shown by the "better" design in Fig. 10.5. Notches and spring-mounted components that ensure location and compensate for wear are another possibility. Sometimes, just by understanding the nature of the difficulty, an innovative way to eliminate the need for adjustment can be found.

Frequently, the need for adjustment occurs when a critical dimension must be established across two or more parts. In general, it is best to avoid this situation by incorporating critical dimensions and orientations into a single part. This avoids all the extra information content required to correctly set the dimension and then maintaining it over the design's life cycle. When

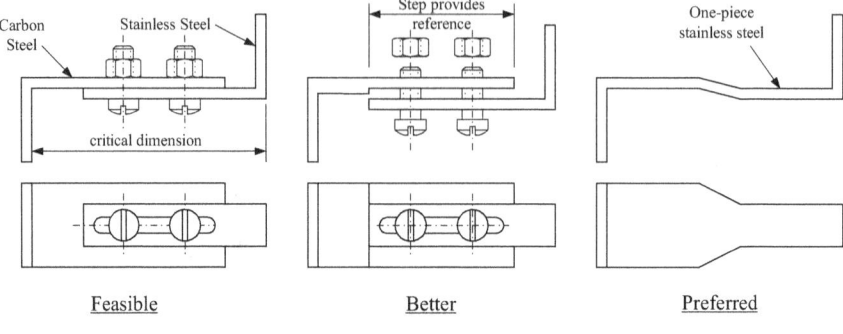

Figure 10.5 Avoiding adjustment reduces assembly cost and quality risk.

this is not possible, use self-aligning and self-adjusting features, avoid multi-hand coordinated procedures, avoid over constraint, use designed-in compliance, and never rely on subjective judgements like "fit at assembly". In almost all cases, the information content and consequent cost and quality risk of adjustment far exceeds the information content of added part features ("better" in Fig. 10.5), material cost ("preferred" in Fig. 10.5), and other design features that eliminate the need for adjustment.

Avoid Separate Operations

Separate operations include all assembly operations other than those directly associated with adding a part, moving to another assembly surface, or performing an adjustment. Examples include mechanical joining processes such as riveting, welding, adhesive bonding, bolt tightening, and so forth. Separate operations inevitably increase assembly information content in the form of extra instructions, material handling, floor space, quality risk, and much more. The ideal assembly is one that requires no separate operations. A layered Z-axis stack (see Fig. 10.2) in which each part is located by a snap-fit or fixture nest and all parts are secured when the cover is assembled is an exemplary example.

Error Proof the Design

Assembly errors generally occur for two reasons: (1) parts are added to the build incorrectly, and (2) the assembly process is unpredictable because of randomness and uncertainty. To error proof the design, these two sources of error must be eliminated.

Design for Assembly: The Second Law of DFM 153

Make Incorrect Assembly Impossible: *Error checking* is the process of determining that the handling, insertion, securing, adjustment, and separate operations have been performed correctly. The information content of this process is greatly reduced when the parts are designed so that they cannot be installed incorrectly. Ways for doing this include the following (Boothroyd, 1994):

- Provide obstructions that will not allow incorrect assembly.
- Design mating features to be asymmetrical so that incorrect assembly is not possible.
- Design parts to be symmetrical so that orientation in unimportant.
- Design parts so that subsequent assembly is impossible if a part is incorrectly installed.
- If necessary, provide "clues" such as matching arrows or colors. Note that this is less desirable than ensuring that incorrect assembly is impossible.
- Provide "keys" and other features on flexible parts such as gaskets to prevent incorrect installation. When possible, avoid flexible components since these can almost always be incorrectly installed.

Eliminate Randomness: Randomness and uncertainty increase the information content of the assembly process. Every effort should be made to eliminate randomness from the design. When that is not entirely possible, design to minimize its affect on the reliability and ease of assembly. Sources of randomness in assembly include undesirable interactions, flexible components, and unrestrained wire harnesses and loose wires.

- Randomness often masks a lack of clarity in the design. Explain all behaviors and phenomena associated with the design. If a behavior cannot be predicted or explained in simple and logical terms, then look for undesirable interactions.

- Design to avoid flexible components when possible. Component flexibility makes positioning and orienting inherently difficult during assembly. Be careful not to increase information content when doing this. For example, nothing is gained if the information content involved in controlling a process-applied gasket is greater than the information content involved in handling an equivalent flexible gasket.

- When flexibility cannot be avoided as in the case of wires and wire harnesses, features should be provided that help ensure position and orientation. Connectors should be mounted in fixed positions to simplify connection. When possible, use rigid circuit boards in place of cables. Locating all connectors at one end of the assembly can greatly simplify assembly operations. When a loose or dangling connector cannot be avoided, it is often helpful to plug the connector into a mating dummy that is in a fixed location and position. This not only facilitates automation; it also ensures that the connector will be located and out of the way until it is needed.

- Great reductions in information content are achievable by planning the layout of wiring harnesses early in the design. In one project, over a mile of wiring was eliminated because all electrical connections were planned before the individual "black boxes" and their connector locations and orientations were designed.

Summary of Key Concepts

> ➢ Next to part count reduction, design for assembly is one of the most effective means for reducing information content of a design.
>
> ➢ Because assembly is an integrative process, it can be the source of many manufacturing and quality problems. These problems can often be avoided or mitigated by planning how the product is to be assembled in the early conceptual stages of design.
>
> ➢ Design for assembly is a systematic application of the eliminate, simplify, and standardize where possible strategy discussed in Chapter 6.

Chapter 11
Piece-Part Design: The Third Law of DFM

A piece-part is an individual component such as a plastic injection molding, casting, or sheet metal stamping that is made using one or a combination of manufacturing processes. The processes used depend on the precision required, the part size, and most importantly, on the quantity of parts to be produced. If the production quantity is high, then near net shape processes that have relatively short cycle times and minimal material waste are preferred since the cost of tooling can be spread out over the large number of parts.

In Chapters 9 and 10, the focus was on part count reduction and design for assembly, which stressed that design decisions should not be based on piece-part cost alone. Rather, the goal should be the *least* number of *simply* shaped parts, produced and assembled using the *least* number of *easily controlled* processing steps. Although not always the case, following this advice will sometimes result in more complex piece-part geometries. Recognizing this, the focus in this chapter is on piece-part cost reduction. Piece-part cost can be reduced in two important ways: (1) minimize information content of the part design, and (2) design to avoid undesirable part geometry/material/manufacturing process interactions.

Information Content

Piece-part manufacture will be inherently easier and less expensive when tooling and process information content is minimized. Toward this end, part features such as radii and hole diameters should be rationalized so that only a few common sizes and shapes are used. In addition, always seek the minimum number of processing steps, ensure that the design specification is well within the process capability, and select parameter values that minimize cycle time and cost. When possible, material alternatives that require surface treatments such as painting, plating, or

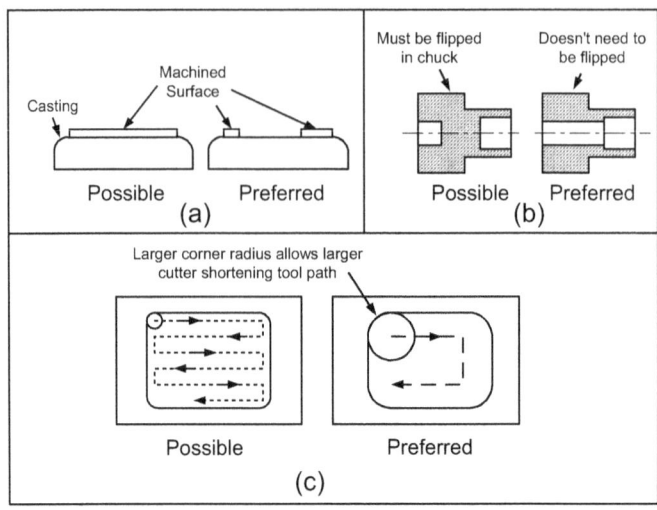

Figure 11.1 Approaches for simplifying the design of a machined part: (a) reduce the volume of material removed, (b) reduce the number of setups, and (c) reduce the tool path length.

polishing should be avoided since the cost of these finishing processes and the equipment and facilities required can far exceed the cost of a more expensive material.

Consider the machining process for example. Recall from Chapter 3 that machining is a material removal process in which the part shape is generated by moving the workpiece and the tool relative to each other. Because the shape information is impressed by tool motion rather than tool geometry, tooling cost is relatively inexpensive making it ideal for low volume production. Also, because it can generate complex and precise geometry, machining is often used as a secondary process to complete parts that are initially formed using shape replicating processes like casting. There are three major process characteristics that must be considered in the design of machined parts. First, the process must be setup. This entails mounting the workpiece in the appropriate work holding fixture and adjusting the machine to perform the desired machining operations. Setup takes time and wastes material, so what ever can be done to simplify or avoid setup is a plus. Secondly, machining processes can involve excessive workpiece handling and manipulation as well as moving the workpiece from machine to machine. As in setups, anything that can be done "by design" to avoid or minimize material handling is a plus. Finally, it is important to design so that both cycle time and tooling cost can be minimized. The following machining design guidelines implement these considerations:

- Reduce the volume of material to be removed (see Fig. 11.1a).
- Drive toward a minimum of machining operations.
- Drive toward a minimum number of machines and setups. For example, design so that all features can be machined from one end (Fig. 11.1b).
- Specify features that allow shorter tool path lengths (Fig. 11.1c).
- When possible, reduce part size to allow the use of smaller, less costly machines and material handling equipment.

Geometry/Process/Material Interactions

It is important to understand that when designing piece-parts, it is not so much the part that is being designed as it is the tooling and process that produces the part that is being designed. Nowhere is this more true than for shape replicating processes like plastic injection molding, casting, and sheet metal forming. The design process for these types of manufacturing processes has been likened by some to the design of a bridge in which a trial bridge is built and a truck is driven onto the bridge causing it to collapse. The reason for collapse is then ascertained and a new, modified bridge is constructed, and the truck is again driven onto the bridge causing it to collapse once again, but for a different reason. This process is repeated until the truck finally makes it across the bridge upon which victory is declared. Although this design process seems totally nonsensical, it is pretty much what is done when part geometry, material, and process interactions are not carefully considered.

In piece-part design, it is essential that the interactions between geometry, material, and process be properly accounted for. When the compatibility of part geometry, fabrication process, and material properties is ensured up-front by design, a short and efficient tool design cycle and low-cost tooling will be the most likely result. This is because the material and detail component design establish, to a large extent, tooling cost, lead time, material cost, cycle time, and process yield (see Fig. 11.2).

Consider the plastic injection molding process for example. This process is a shape replicating manufacturing process that is widely used to produce large quantities of production parts. To produce a part using this process, polymer material is first heated to form a "melt". The melt is then forced under pressure to flow through a system of channels (runners) and gates into one or more mold cavities, where the melt solidifies to form the part(s). The parts (and runners in some cases) are then mechanically ejected from the mold cavities and a new cycle is initiated. In plastic injection molding, there are five major process considerations that must be understood and properly

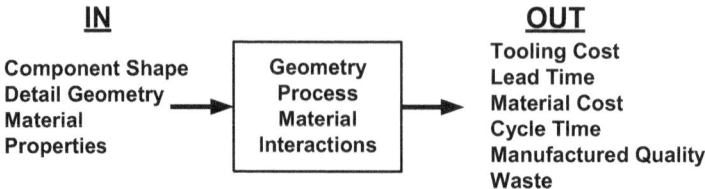

Figure 11.2 Compatibility between the design and process determine cost.

designed for: (1) material shrinkage, (2) the need for draft, (3) gating location, (4) parting line design, and (5) undercuts.

Shrinkage: Material shrinkage occurs due to thermal contraction of the material as the part cools. Because thin sections cool faster that thick sections, shrinkage can cause a variety of part defects (see Fig. 11.3a). In addition, because "cooler" regions are more rigid than "hotter" regions, stress can develop due to constraints that form as the part contracts onto the mold cores (male parts of the mold). These stresses are "frozen-in" as the part solidifies resulting in undesirable residual stress that degrade the part's strength and integrity. To avoid shrinkage defects and residual stress, the part must be designed using a uniform wall thickness and generous radii (see Fig. 11.3b). By using a uniform wall thickness, often referred to as the *nominal* wall, thermal mass, and therefore shrinkage is the same everywhere in the part. Sharp corners, especially inside corners, should be avoided because they cause severe molded-in stresses as the material shrinks onto the core. Also, tight radii cause poor flow patterns and increased tool wear.

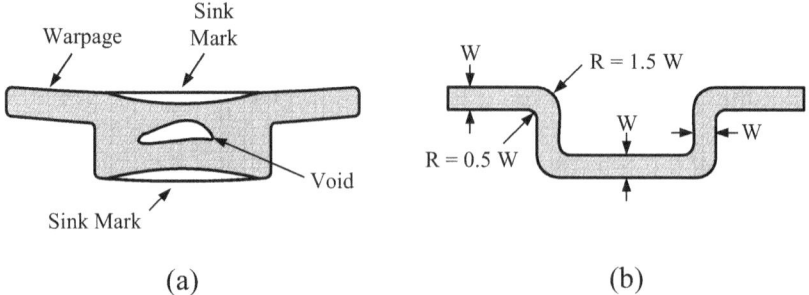

Figure 11.3 (a) Shrinkage defects occur due to variations in thermal mass in different regions of the plastic injection molded part. (b) A uniform wall thickness (nominal wall) combined with generous fillets and radii help avoid shrinkage problems.

Piece-Part Cost Reduction: The Third Law of DFM

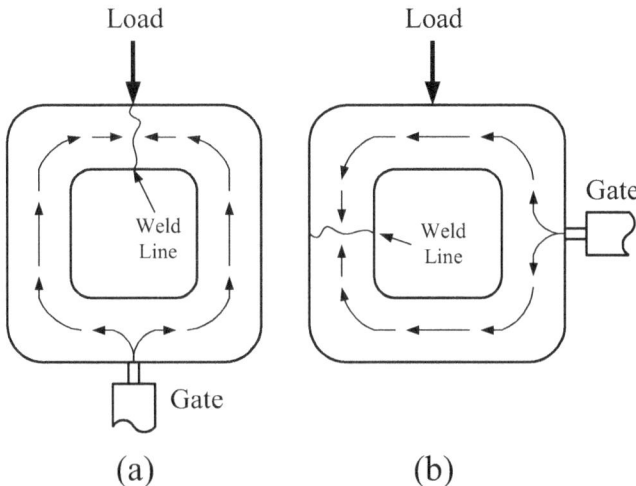

Figure 11.4 Welds lines form when plastic melt flows meet. (a) This gate location causes a weld line to form in a critical region of the part. (b) This gate location causes the weld line to form in a less critical region.

Draft: Shrinkage also causes the part to "grip" cores and other tooling components. This, along with friction between the part and the mold walls, makes it difficult to eject the molded part from the mold cavity. To offset this affect, surfaces that are perpendicular to the parting plane must be tapered so that clearance between the part and the mold occurs as soon as the mold begins to open. Commonly called *draft*, this need to taper various part surfaces can compromise appearance and functionality of the molded component and must be carefully considered during design.

Gate Location: While need for a nominal wall and draft are dictated by material shrinkage effects, "gate location" is primarily dictated by process effects. Runner channels in the mold conduct the melt flow to the mold cavity. The *gate* connects the runner channel to the mold cavity and is shaped and sized to meter the flow into the mold cavity at a desired rate. Because gate location determines where and how the melt enters and flows through the mold cavity, it has a direct bearing on the quality of the molded part. As shown in Fig. 11.4, gate location can affect the strength of the part. It can also affect the orientation of glass fibers or other fillers. Gates may leave burrs or sharp protrusions on an otherwise smooth edge or surface. These can, in turn, interfere with automated feeding devices such as bowl feeders. Importantly, if these rough protrusions are not properly located or otherwise removed by additional processing steps or increased tool complexity, gates can detract from the quality and appearance of the component.

Parting Line: When a mold closes, the core and cavity meet to create a closed air space into which the plastic melt is injected. From the inside, the mating junction between the mold halves appears as a line. This line also appears on the molded part and is called the *parting line*. The selection of the parting line depends on the shape of the part, method of part ejection, type of mold, aesthetic considerations, post-molding operations, inserts, venting, wall thickness, number of mold cavities, and the location and type of gating. Selection of the parting line has far reaching consequences because it impacts many aspects of the part, such as the surfaces that must have draft, the surfaces upon which ejector pins can act, where and how the part is gated and the mold vented, the tolerances that can be achieved, and the amount of mold clamping force required to hold the mold halves together as the melt is injected under pressure into the mold cavity or cavities.

Undercuts: Geometrical features of a component design that prevent mold opening and/or part ejection are known as *undercuts*. Moving cores, side actions (camming), unscrewing devices, and other "tricks of the trade" are commonly employed by the tool designer to overcome the effect of undercuts. It is also possible to avoid undercuts by performing post-molding secondary operations such as drilling or machining. In some cases, it is possible to eliminate undercuts by designing the part and selecting the parting line such that functional and appearance requirements are satisfied without the need for undercuts.

If the above geometry/process/material interactions are not properly understood and accounted for, tool cost, material cost, cycle time, and process yield are all likely to be adversely affected. Some design engineers have worked extensively with one or more processes throughout their career and possess extensive process knowledge. Such engineers are relatively rare, however. Most design engineers are not process experts. In these cases, it is extremely important that the design engineer have ready access to process expertise. This generally requires that the design engineer work closely with manufacturing engineers, tool designers, and others who know the process much better. Without the availability of specialized knowledge, it is highly likely that an unnecessarily expensive part will be designed.

Improving the Piece-Part Design Process

In Chapter 4, we discussed the iterative nature of design. If we look at the conventional approach to component design, we see that a multitude of iterations can occur during the component development (see Fig. 11.5). First the design engineer goes through the iterative design process to specify the component geometry. This geometry is then passed on to the manufacturing

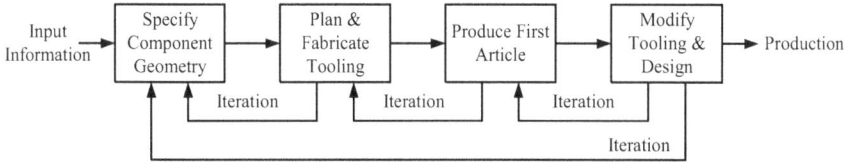

Figure 11.5 Conventional component design process.

or tooling engineer who repeats the iterative design process to specify the tool and process design. Problems discovered during this stage generate additional iterations if component geometry changes are called for. Additional iterations to the component geometry and tool design may also be required during tool fabrication and preparation for production of the first article. Finally, iterative changes to the tooling and perhaps the component geometry may be necessary to fix problems with the first article and to "tweak" the design to meet production requirements.

This excessive design iteration is driven by the tight coupling between geometry, material, and process considerations. It is this iteration that significantly increases design cost and design time of complex parts. Most importantly, design iterations performed late in the process can lead to suboptimal design. The result is a component that falls short of cost and performance targets. Such designs place the whole project in jeopardy.

How can the conventional component design process be improved? Experience has shown that there are several steps that can be taken to improve the situation by cutting down significantly on the many often avoidable iterations that commonly occur. These include the following proven suggestions.

1. Design the component geometry and fabrication process as a coordinated system in one concurrent process. Consider geometry, material, and process interactions and design related cost drivers from the beginning of the design process.

2. Develop a thorough understanding of all customer needs including downstream processing constraints before beginning the component design.

3. Focus on creating an acceptable initial design. By spending the time "up front" required to create the best possible initial design, many lengthy analyze-redesign iterations can be avoided. The evaluation phase should confirm the design rather than create it.

4. Use manufacturing process simulation software and other modern computer-aided analysis and inspection methods to quickly optimize the design.

5. Develop a consistent, well-defined "science base" for component design by developing design guidelines and structured methodologies for each core manufacturing process or method used by the firm.

The goal of these suggestions is to reduce cost by shortening the component and tooling design cycles, ensuring an acceptable first article, and helping to ensure that the full complement of design and process knowledge is available when it is needed. A few years ago, I became involved in a project aiming to better understand the science of aluminum sand casting. My interest was in developing a deeper understanding of how the design of individual part features interact with the tooling and sand-casting process used to make it. As an experimental exercise, I designed a relatively complicated aluminum casting that required several intricate cores and posed some other challenging problems of design. I then sat down with a very experienced sand-casting tooling designer and a foundry engineer and showed them the part I had designed. They quickly showed me how two cores could be eliminated without compromising the design in any way. In addition, they suggested a simplified parting line that hadn't occurred to me and that would result in a much lower cost tooling design that could be implemented in a much shorter design cycle. For me, this was an "eye-opening" lesson in why these suggestions are so important. High bandwidth communication between all relevant knowledge sources is the secret.

A Meeting-Based Approach

One of the best ways to ensure the ready availability of needed knowledge is to use a structured team approach. In a *structured team approach*, the overall problem of design is broken down into a series of sequential, easier to perform steps that proceed from the general to the specific. Often, each step in the process can be further subdivided into steps to create a hierarchy of structured methodologies.

A simple meeting-based component design process (see Fig. 11.6) illustrates the structured team approach. With the right team member attitude, meetings can be a highly effective means of communication. This is especially true when the meeting brings differing expertise together. A meeting-based approach also recognizes that not all team members can be available on a continuous basis, especially when the team involves members from outside the firm such as suppliers and tooling houses.

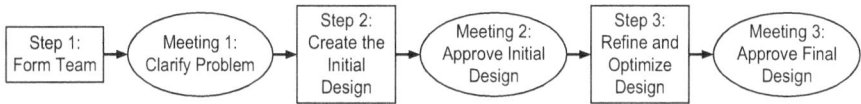

Figure 11.6 Simple "meeting based" structured team approach.

The purpose of the team meetings is to establish design direction, make key design decisions that require input and consensus from all team members, make sure that all process constraints and requirements are being properly considered, and resolve conflicts and impediments to the proposed design. Because of the importance of these meetings, it is imperative that all members of the team be present at each meeting. The outcome of each meeting is a set of action steps to be implemented by individual team members. In this way, all team members are kept informed and participate in the design decision-making process. At the same time, the actual detail work of creating the design is delegated to specific team members according to the skills and knowledge required. In the following, how a component would be designed is imagined by briefly discussing each step in the methodology.

Step 1: Form Team

This is the pivotal first step in the methodology. Unless the arrangement is formalized in some way, it can be difficult to get effective collaboration between the design and manufacturing engineer early in the component design process, especially before the component configuration and geometry are defined. By being formally assigned to a team, each individual team member takes personal responsibility for the design from the beginning. This encourages and facilitates the kind of collaborative attitude that is essential for the structured team approach to succeed.

All "stakeholders" who have an interest in the component should be represented on the team. A typical team might include a design engineer, a process engineer, a tooling design engineer, and perhaps one or more specialists who are familiar with process simulation software, finite element analysis, fracture mechanics, non-destructive evaluation (NDE) techniques, and so forth. In addition, if the component is to be purchased from an outside supplier, it is essential that the supplier be properly represented on the team. This not only helps ensure that all customer and processing needs are appropriately considered, it also makes it possible to rapidly negotiate changes to the design specification when necessary, and to quickly assess cost consequences of design decisions.

Meeting 1: Clarify the Design Problem

Clarifying the problem consists of developing a general understanding of the cost, performance, and manufacturing goals and constraints of the design. A typical agenda for this meeting might include the following:

1. Review product background.
2. Customer requirements and design objectives.
3. Expected annual production volumes, target costs and lead times.
4. Geometry concepts and alternatives.
5. Material and processing options.
6. Production facility and secondary processing locations.
7. Potential geometry/material/process interactions.
8. Developing a preliminary configuration design.
9. Making assignments to team members to create the initial design.

As mentioned previously, it is essential that all team members be present at each team meeting. For example, although the analyst may not be actively involved with the design before the component geometry is fully defined, it is extremely important that he or she participate in the early design decisions that lead to the proposed geometry. In this way, the analyst knows all the needs of the design problem and is familiar with the reasoning behind the geometry that will eventually be analyzed and optimized.

Step 2: Create the Initial Design

The initial design establishes the detail layout of the component geometry and the tooling and/or process concept. It includes the configuration and parametric design of the part together with tooling and processing information such as parting line selection, gating location, tooling surfaces, datums, and so forth. *Configuration design* involves the determination of what features such as walls, holes, ribs, etc. will be present and how these features will be connected to provide the desired form, fit, and function. *Parametric design* involves the determination of dimensions, tolerances, and exact material specifications needed to meet durability, stiffness, natural frequency, and other functional and performance targets.

As a general approach, a preliminary configuration might be proposed in Meeting 1 by the team. Using this as a starting point, the design engineer and manufacturing engineer work together to develop the details of the configuration design, seeking input and consensus from various team members as necessary. Once the initial component configuration has been firmed up, a preliminary parametric design would be performed. The goal of this task is to quickly determine section dimensions, secondary processing requirements, and material property requirements. Once the approximate parametric design is complete, the overall design is evaluated and modified to minimize cost. This is easy and straightforward to perform with a minimum of analysis and iteration because the component geometry and tooling concept have been conceived and developed as a coordinated system with input from all team members.

Meeting 2: Refine and Approve the Initial Design

The goal of this meeting is to react as a team to the initial design and to make any adjustments or modifications deemed necessary by general team consensus. This is the time when all design and processing issues should be discussed and resolved. If there are significant impediments to the design as proposed, these should be resolved before proceeding to Step 3. A typical agenda for this meeting might include the following:

1. Review the initial design.

2. Identify impediments, potential undesirable interactions and performance and processing concerns.

3. Discuss all design-related costs to ensure that the best component geometry from a total cost standpoint has been identified.

4. Make assignments to individual team members to work out solutions to various impediments and schedule a follow-on meeting.

5. If no impediments are identified, approve the initial design and make assignments to team members to refine and optimize the design.

Step 3: Refine and Optimize the Design

Once the team is confident that the initial component geometry and tooling design is the best solution possible, the effort required to optimize details of the component geometry and tooling concept by computer analysis can be justified. The goal of this step is therefore to computer model the design and iteratively improve it until all aspects have been appropriately optimized. The amount of effort expended will depend on how important it is to optimize the component. For example, if weight and/or material cost is critical, extensive optimization effort can be justified. Similarly, if safety is an important issue, comprehensive analysis to ensure acceptable fatigue life and reliable detection of flaws can be justified. The key to this step is to start with a component geometry that is close to the optimum. This will minimize the number of iterations required to converge to a final design.

Meeting 3: Approve the Final Design

The result of Step 3 will be a fully specified component design including the detailed component geometry, tooling design, and fixture design. In addition, the finished component design including machining, heat treating, and so forth will be fully specified. The purpose of Meeting 3 is to formally review the finished design as a team and approve the design for release to manufacturing. When the structured team approach is performed properly, the final design will almost always be approved. However, if the team decides that the design is not ready to be released, then appropriate action plans for correcting design deficiencies must be developed and implemented. One or more follow-on meetings may then be required before the design is released.

Although painful at times, not releasing the part until it is ready helps ensure a minimum number of tooling changes and "tweaks" and, in the long run, is the most cost-effective policy. By strictly adhering to this policy, the component design should proceed quickly and smoothly to first article and production with little or no modification. When this is the case, the team knows that it has done its job well.

Summary of Key Concepts

- ➢ The goals of piece-part design are (1) a rapid and efficient design process, (2) a functionally acceptable part that can be manufactured for the lowest possible total cost and highest possible quality, and (3) a design that transitions smoothly into full scale production.

- ➢ High quality communication channels between all stakeholders are essential to help ensure that part geometry, material, and manufacturing process interactions are properly considered and accounted for at all stages of piece-part design.

- ➢ Often, structured methods, such as the team-based approach suggested in this chapter, offer an effective way to ensure that all needed part geometry, material, and process knowledge is available when key design decisions are made.

Chapter 12
The Geometric Layout Improvement Method

Structured DFM methods are systematic design and analysis procedures that help facilitate, promote, and ensure good design for manufacture. This chapter presents an easy-to-use structured approach for systematically improving the geometric layout of a design. Recall that *geometric layout* refers to the way a design is divided into subassemblies and separate parts. Because it determines the number and complexity of designed parts, for many designs, geometric layout, more than any other early design decision helps to maximize total design value.

Geometric layout improvement seeks to achieve the next plateau of cost reduction by reducing manufacturing and assembly complexity and quality risk. The method is focused on optimizing the geometric layout using the laws of DFM discussed in Chapters 9 through 11. The goal is to eliminate as many parts as practical and then to design the parts that remain to be easy to manufacture and assemble. The method works by minimizing information content using the eliminate, simplify, and standardize where possible strategies to both guide and test for good design for manufacture.

The method applies the minimum part count assessment presented in Chapter 9 together with the DFA guidelines discussed in Chapter 10. It is implemented using the DFM design-analyze-redesign process suggested in Chapter 5. It begins with an initial design and continues through an iterative process of design improvement until an acceptable design for manufacture and assembly is achieved. The method is qualitative rather than quantitative. This makes it simple to use by avoiding the need to perform lengthy numerical calculations. The worksheet can easily be implemented as a simple computer spreadsheet.

The iterative design-analyze-redesign process is performed in four steps:

1. **Gather Information:** Obtain the best available information.
   - Engineering drawings (solid model, exploded view, etc.).
   - An existing version of the design; a working prototype; a proof of concept model.
2. **Analyze:** Determine geometric layout improvement opportunities.
   - Take the design apart (or imagine how this might be done). If the design contains sub-assemblies, treat these, at first, as "parts" and then analyze them later as assemblies.
   - Begin reassembling the design in the reverse order from which it was disassembled. Imagine that the build can be reoriented so that each part is added <u>top-down</u> using <u>one-hand</u> and keep track of each reorientation. As each part is added to the assembly and regardless of practical or functional limitations, answer each of the following questions:

     (1) Does the part move relative to other parts?
     (2) Must the part be made of a different material?
     (3) Does the part need to be separate for assembly or service?

   - If the part receives an answer of "Yes" to *any* of these questions, then the part is a theoretical part and *probably* cannot be eliminated. If the answer to all three questions is "No", then the part is identified as a candidate for elimination (CFE).
   - As each part is added to the build, analyze it for opportunities to eliminate information content by avoiding reorientations and improving ease of handling and assembly.
   - Record improvement opportunities using the "Geometric Layout Improvement Method Worksheet" shown in Fig. 12.1. Alternatively, use an equivalent computer "spreadsheet".
3. **Redesign:** Implement improvement opportunities identified in Step 2. To maximize the possibility of identifying a "best" redesign, develop several alternative redesigns that range from simple minor changes to speculative (and possibly risky), but desirable "far-out" ideas. Eliminate information content by eliminating parts and designing those that remain to be easy to assemble.
4. **Winnow, Refine, Optimize:** Develop improved redesign concepts by eliminating undesirable features and combining desirable features of the redesign alternatives proposed in Step 3.

The Geometric Layout Improvement Method

GEOMETRIC LAYOUT IMPROVEMENT METHOD WORKSHEET

Name of Design _____   Sheet _____ of _____

Part Name	Qty	Motion	Material	Assembly/Service	CFE	Notes

$\sum Parts =$ _____ $\sum CFE =$ _____

$$Count\ Efficiency = \frac{\sum Parts - \sum CFE}{\sum Parts} \times 100\%$$

Figure 12.1 Geometric Layout Improvement Method worksheet.

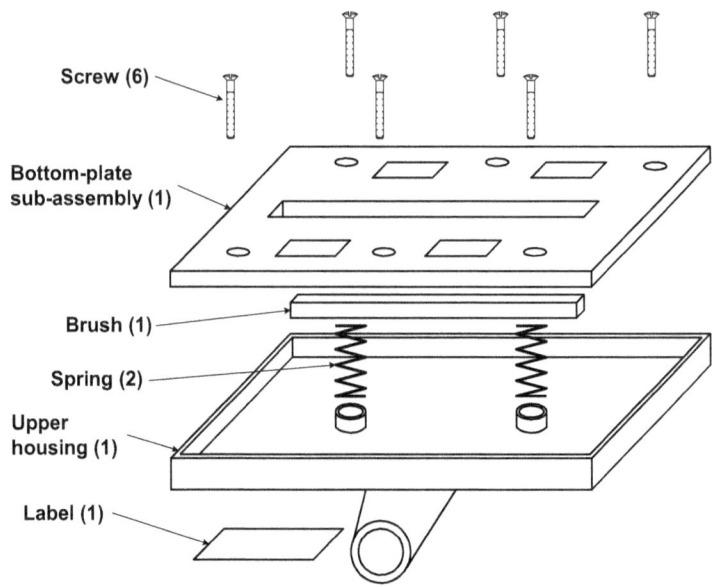

Figure 12.2 Exploded view of a vacuum cleaner attachment.

Illustrative Example 12.1

Improve the geometric layout of the vacuum cleaner attachment shown in Fig. 12.2. Six screws hold the bottom plate subassembly to the upper housing. The bottom plate subassembly, which consists of the plastic injection molded bottom plate together with several pressed in pin-mounted rollers that facilitate movement of the attachment along the floor surface being vacuumed, is treated initially as a single part. An asymmetrical molded nylon brush is sandwiched between the bottom plate sub-assembly and the upper housing. The brush extends through a slot in the bottom plate and is held in contact with the floor by two metal helical springs that fit into circular nests molded in the upper housing. The vacuum attachment is disassembled by removing (1) six metal screws, (2) one bottom plate assembly, (3) one brush, (4) two springs, and (5) one label. The upper housing is the last remaining part. This part is designated as the "base" part or "starting" part.

Analysis: Starting with the upper housing (the base part), reassemble in the reverse order. Analyze each part as it is added to the build by completing one line on the worksheet after it has been added "top-down" to the build using "one hand". The part count is recorded in the quantity column and any assembly difficulties are noted. The three critical questions are asked of each

The Geometric Layout Improvement Method

GEOMETRIC LAYOUT IMPROVEMENT METHOD WORKSHEET

Name of Design ____Vacuum Attachment____   Sheet __1__ of __1__

Part Name	Qty	Motion	Material	Assembly/Service	CFE	Notes
Upper Housing	1	No	No	Yes	0	
Label	1	No	No	No	1	Sticky; hard to position
Reorientation	1	—	—	—	1	
Spring	2	No	No	No	2	Tangle; hard to handle
Brush	1	Yes	No	No	0	Orientation not obvious; holding required
Bottom Plate Sub-assembly	1	No	No	Yes	0	Hard to guide brush through slot
Screw	6	No	No	No	6	

$\sum Parts = $ __13__ $\sum CFE = $ __10__

$$Count\ Efficiency = \frac{\sum Parts - \sum CFE}{\sum Parts} = \frac{13-10}{13} \times 100 = 23\%$$

Figure 12.3 The worksheet analysis for the vacuum attachment.

part and if the answer is "no" for all three questions, the part is identified as a candidate for elimination (CFE) and the part count that is recorded in the the quantity column is also recorded in the CFE column. Separate operations, such as the reorientation required after the label is added, are recorded on a separate line of the work sheet and are always CFE's. The worksheet and analysis for the vacuum attachment is shown in Fig. 12.3. Observations and insights gained from the analysis include the following:

- Since the base part is the first part in the assembly, the motion and material questions both receive "no" answers. The assembly/service question is a "yes" however, because a base part is needed to begin the assembly.

- As noted, the label is difficult to handle because it is sticky on one side. It is hard to position because it is flexible.

- The reorientation is necessary so that the springs can be added "top-down". The method assumes that the build remains intact regardless of orientation.
- The coil springs are difficult to handle because they tend to tangle and nest. A "no" answer is assigned to the motion question because one end of the spring is stationary.
- The brush is difficult to handle because it is asymmetric, and the correct orientation is not obvious. It is difficult to insert because the springs need to be held in position.
- The bottom plate is difficult to insert because there is no taper or other feature to guide the brush bristles through the slot. The assembly/service question is answered "yes" for two reasons: (1) the part is a sub-assembly and (2) it must be separate to allow assembly of the springs and brush.
- Separate fasteners will always receive "no" answers to the critical questions and are always treated as CFE's.

Re-Design: The geometric layout is improved by seeking creative ways to harvest the improvement opportunities identified in the analysis step. The goal is to eliminate CFE's and to design the parts that remain for ease of assembly and manufacture. To fully explore all improvement possibilities, it is recommended that the redesign proposals developed include a "practical", a "stretch", and a "radical" redesign. A *practical* redesign is one that can be implemented easily without excessive tooling modifications or other major changes. A *stretch* redesign, on the other hand, could involve significant development effort and cost. A *radical* redesign is a stretch redesign that involves a major change such as switching to a totally different physical concept or to a new material or method of manufacture. To illustrate, consider the following vacuum cleaner attachment redesigns.

Concept A, Practical: (1) Eliminate the label by integrating it into the upper housing either as a "hot stamping" or as a molded-in feature. (2) Replace the screws with snap fittings. This can be done by making relatively simple changes to existing tooling. (3) Mold-in a tapered guiding feature on the bottom plate to orient the brush and guide the bristles through the slot. Again, this requires simple modification of the existing tool. (4) Modify tool to provide snap-fittings in the upper housing that retain the brush thereby avoiding the current instability and holding requirement. (5) Assemble as a "Z-axis" stack without the need to reorient the build.

The Geometric Layout Improvement Method

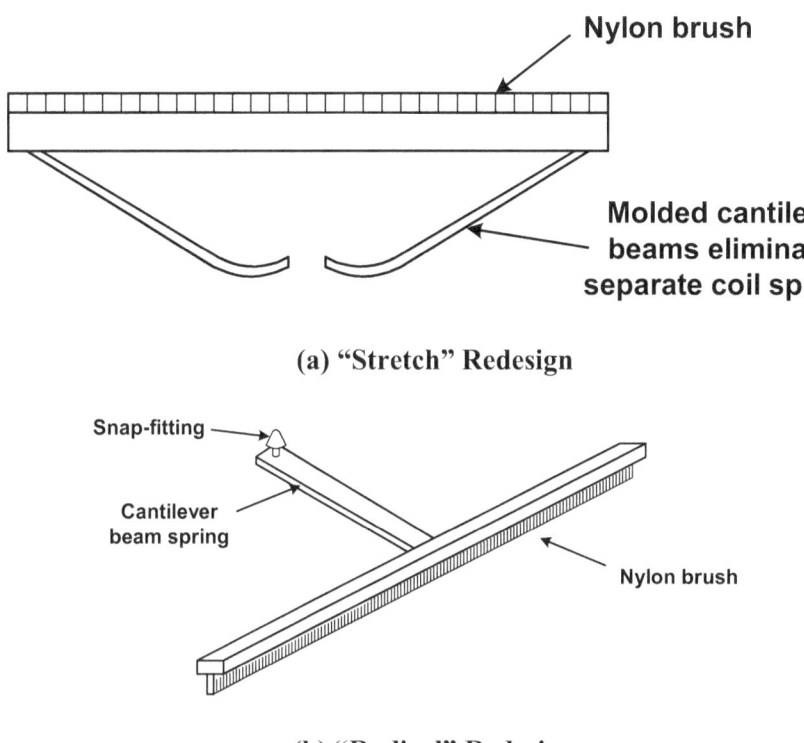

(a) "Stretch" Redesign

(b) "Radical" Redesign

Figure 12.4 Vacuum attachment redesign concepts.

Concept B, Stretch: Same as concept A plus eliminate the separate coil springs by integrating the spring function into the brush. One approach might be to incorporate cantilever springs on the brush as shown in Fig. 12.4a. This will most likely require new tooling.

Concept C, Radical: (1) Combine the bottom plate and upper housing into a single part. (2) Integrate the brush and coil springs into a single part that snap-fits into the upper housing and that is exactly constrained by molded-in features. One possibility is shown in Fig. 12.4b.

Winnow, Refine, and Optimize: In the "winnow, refine, and optimize" step, the performance and cost implications of each redesign proposal are evaluated in view of customer needs as well as development time and budget constraints. "Hybrid" concepts that combine the best features of each design proposal are also developed and evaluated. For example, a "hybrid" concept D can be created by modifying concept B to use the simpler spring/brush design from concept C.

Table 12.1 Comparison of redesign concepts A, B, and C.

Redesign Concept	Part Count	Parts Eliminated	Sub-Ass'y Eliminated	Count Efficiency
Original Design	12	---	---	23 %
Concept A	5	7	0	60 %
Concept B	4	8	0	100 %
Concept C	2	10	1	100 %

Discussion of Results: Redesign concepts A, B, and C are compared as shown in Table 12.1. This comparison illustrates that significant part count reduction is possible using a practical redesign (concept A). It also shows that further part count reduction tends to be accompanied with increased tooling cost and implementation complexity (concepts B and C). At the same time, reduction in information content can be extraordinary, usually resulting in cost savings that far exceed the increased tooling cost of the more complicated parts. Concept C, for example, eliminates all the information content associated with the bottom plate sub-assembly and replaces a 12-part design (not counting the parts in the sub-assembly) with one having just two separate parts. It should be noted that, in creating concept C, the team had to reevaluate the "yes" answer to the "assembly/service" question for the bottom plate assembly. This illustrates a subtle pitfall associated with the three critical questions. Often a "yes" to one of the questions is the result of the particular geometric layout being considered and can be converted into a "no" by a different geometric layout approach as illustrated by concept C. Therefore, answering "yes" to a critical question should always be carefully considered and all "yes" answers should be challenged during the redesign step. This is especially true for the "assembly/service" question. It should also be noted that there are many possible metrics that can be used to measure the "quality" of the design with respect to information content. The "count efficiency", which compares theoretical part and separate operation count to the actual count, acts as a simple to calculate measure of the information content. As illustrated by concepts B and C, however, it is also an imperfect measure because it depends on how the three critical questions are answered. Therefore, in practice, it is important to focus on reducing information content of the design rather than on improving metrics such as the count efficiency.

The Geometric Layout Improvement Method

Summary of Key Concepts

> ➤ The first geometric layout is seldom the one that is most manufacturing and assembly friendly. To identify an improved design for manufacture, it is best to analyze the starting design for information content and then iteratively redesign to reduce information content.

> ➤ Almost all designs, existing or new, can be cost reduced by analyzing the design using the geometric layout improvement method. In many cases, simple, low cost improvements that have minimal impact on tooling and manufacturing method are possible.

> ➤ When a part receives the answer "Yes" to one of the critical questions, the natural assumption is that the part is a theoretical part and cannot be eliminated. This is not always the case. For example, parts that undergo small relative motions can sometimes be eliminated by using a "flexure" or "living hinge". Often it is the physical concept or the geometric layout itself that is forcing the "Yes" answer. Look for an alternative concept that avoids the "Yes". Always challenge theoretical parts.

Chapter 13
Robust Design: Combating Variation and Change

Variability, randomness, uncertainty, unpredictability, and change are all enemies of manufacturing. Extra steps and activities are required to correct problems, rework and repair defective product, and process warranty claims and customer complaints. The result of hard-to-control variation and randomness is poor quality, unnecessary manufacturing cost, product unreliability, and ultimately, customer dissatisfaction and loss of sales. The effect of change can be even more devastating. In addition to disrupting marketing and business operations, design and technological change can result in the obsoletion of entire production lines and manufacturing facilities. Disarming these threats and preventing the costly consequences that they create through product design is the subject of this chapter.

Design decisions, made in the early stages of design, can often reduce, and possibly eliminate the cost associated with variability, uncertainty, and change by making the product inherently robust. The term "robust" implies a product designed to perform its intended function no matter what the circumstances. A robust design is one that is insensitive to and/or tolerant of change and variation in operating conditions, manufacturing conditions, and business conditions. A "robust" design has the following characteristic:

- The product functions and performs the same regardless of variation in day-to-day operating conditions such as temperature, humidity, input voltage, severity of loading, type of use, and so forth.

- The product functions and performs the same regardless of variations or changes that occur as the product is used and as it ages and deteriorates over time. Examples of changes that might occur over the products life include material fatigue, component wear, and environmentally induced deterioration such as corrosion.

- Product to product variation due to hard-to-control manufacturing variation does not affect product performance or functionality.

- The product functions and performs the same regardless of changes in suppliers and business conditions.

- Product characteristics such as features, performance specifications, capabilities, and styling are easily adapted to changing market conditions with minimal investment and timing consequences.

A truly robust product is one that has been intentionally designed to be robust. The best time to do "robust design" is in the early conceptual stages of design when there is the most latitude to explore alternative possibilities and therefore the greatest opportunity to address all aspects of robustness. In addition, early consideration of robustness greatly increases the probability of identifying and selecting the inherently best product and process concept.

Robust design is accomplished in a variety of ways ranging from "platform" design strategies to detail assembly and component design. Some of the more effective and successful of these are discussed as follows.

Develop an Inherently Robust Design Concept

One of the surest ways to achieve a truly "robust" design is to begin with a design concept that is inherently insensitive to hard-to-control variation. Recall the independence axiom from Chapter 7: *Maintain the independence of functional requirements*. Design concepts that are based on this axiom are, by definition, inherently robust. By separating or decoupling functions, changes or variations that affect one functional requirement will not affect or ripple to other requirements. The radar antenna support example discussed in in Chapter 2 is illustrative of this important idea.

The seal design concepts shown in Fig. 13.1 further exemplifies this point. In design concept A, relative motion between the seal and rotor surface generates frictional heat that causes the rotor diameter to expand producing undesirable seal wear. To make matters worse, the rotor expansion causes the seal to press even harder generating additional heat and further increasing rotor expansion. In design concept B, rotor expansion decreases the seal pressure thereby decreasing the amount of heat generated and allowing the rotor to settle into a stable, self-regulating diameter. Design B is clearly preferable because it prevents the thermal instability from ever occurring. This is what is meant by "inherent" robustness.

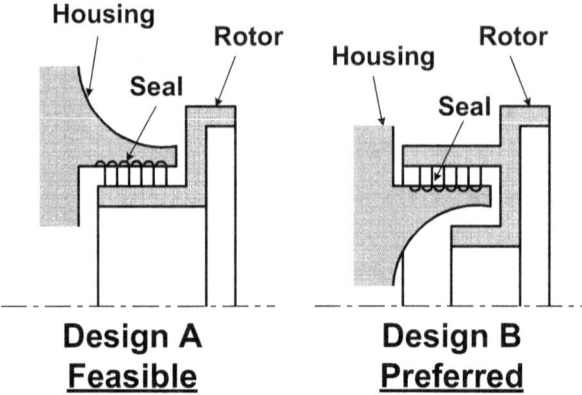

Figure 13.1 Alternative rotor seal designs. Design A is unstable because thermal expansion increases seal pressure on the rotor. Design B, on the other hand, is stable because thermal expansion decreases seal pressure.

Design for Change

Change has become an increasingly important consideration in a product's lifecycle. Customer needs and perceptions change, new product innovations and technology breakthroughs occur regularly, competition is constantly challenging and pushing current products, and new materials and processes are continually emerging. Changing market needs and technology often initiates a design cycle. Change continues to occur during the design process due to iteration and uncertainty. In addition, in some cases, it may occur because a new product introduction generates new needs and applications. Like variation, change creates chaos and non-competitiveness when improperly managed or inadvertently overlooked.

Consider, for example, the design of an electrically driven machine. Suppose that when first designed, the electric motor used to drive the machine was purchased from a preferred supplier. Later, a new purchasing manager is hired who finds an offshore source for a comparable motor that can be purchased in the quantities needed for a much lower price. Unfortunately, the mounting hole pattern in totally different. This is no big problem if all that is needed to accommodate the new motor is to drill a new hole pattern. It becomes a more serious problem if the motor is installed robotically because, to accommodate the new motor, the end of arm tooling must be redesigned. If the motor is mounted to a casting or other hard to change part, then the expensive redesign effort required may rule out consideration of the lower cost motor altogether.

As a second example, consider a product that is initially designed as a single purpose device intended to satisfy an unmet market need. As the product becomes available in the market, the company soon discovers that there is a whole range of unmet needs and that the solution they have provided is in demand. Suddenly, what started out as a single purpose product has evolved into a family of products, each requiring specially designed and manufactured components, different production lines, and different support, installation, and servicing needs.

What these two examples and many more like them illustrate is that whether due to new product technology, new manufacturing technology, or new business conditions, change itself represents an undesirable interaction between present and future functional requirements. There is a variety of ways to "decouple" this undesirable interaction. For example, in the electric motor example, the unique motor geometry can be effectively isolated from the rest of the product and/or end of arm tooling by using an "adaptor plate" that interfaces the motor to the product and manufacturing environment.

The objective of design for change is two-fold: (1) make the inevitability of change compatible with the need for a stable and disruption free manufacturing environment, and (2) minimize capital investment and timing consequences that may be incurred due to inevitable and necessary design change. Hard to control factors include:

- Changing day-to-day production, customer, and market needs.
- Competitive pressure.
- Availability of new technologies and materials.
- Business and manufacturing uncertainties.
- Change due to continuous improvement of the product and process.

Several strategies for making the product/process robust against change are possible. One of the most effective of these is standardization. The key in using standardization effectively is to identify the right standardization principle. Agreeing early to design a machined part so that all features can be obtained by machining on only three orthogonal axis makes the design readily adaptable for manufacture on a flexible machining center. Use of a standardized base component for all variants of a product effectively decouples the material handling system from the product variant being assembled in a flexible assembly process. Reverse engineering of many successful integrated product and process designs reveals similar underlying standardization principles or other "decoupling" strategies.

The trick is to anticipate change in the early stages of design and design-in features that accommodate it. To do this, the team needs to look five or six product generations into the future and anticipate changes that are likely to occur. Once identified, the team can consciously plan for change in the design. Like the electric motor example, it is a huge "win" if all the anticipated change can be isolated into a single part or module or "chunk". When this is possible, accommodating change, no matter what or how far-reaching and complicated it is, can be done quickly and with a minimum of cost impact and disruption.

Building on the design-analyze-redesign DFM approach, one way to anticipate and design for change is to use a simple three-step procedure.

Step 1: Evaluate the proposed product concept and manufacturing plan for possible changes that may occur over the lifetime of the product.

- What product and applicable technology trends and changes are likely? How might customer needs or functional requirements change as a result? How would the product be affected by these changes or new developments? How would such change, if it occurred, be accommodated by the design?

- What business driven manufacturing changes are likely to occur? How is applicable manufacturing technology trending and how could such changes affect the firm's manufacturing environment? How would these changes affect production of the product? How would they be accommodated by the product/process design?

- What product or model variations are planned? How does the product design and manufacturing system accommodate these changes? What new variations or product models could be introduced in the future? How would these changes impact the current product design and manufacturing plan? How would they be accommodated?

Step 2: Analyze the results of the evaluation and develop ideas and approaches for accommodating anticipated change. Seek coordinated product design and production system concepts that "decouple" change from the method of manufacture and isolate the effects of change in planned ways.

Step 3: Redesign the product concept and manufacturing plan according to the ideas adopted in Step 2 and reanalyze. Iterate until satisfied.

In following this procedure, each question in Step 1 should be answered in as much detail as possible, given the state of product and process detail available. If the product concept and manufacturing plan are nearing completion, then the procedure will help identify vulnerabilities. If the design is in its early stages, then the procedure will help improve design objectives and identify issues and concerns. The whole point of the exercise is to convert unanticipated change into a comprehensive plan that accounts for and accommodates change.

Design Using Proven Design Principles

Design principles and best practices that result in good design often also produce beneficial improvements in design robustness. The principles should be used to guide geometric layout decisions even when robustness is not a primary concern.

Design Parts to Have Uniform **Strength.** This principle seeks to ensure, through careful selection of materials and shapes, that each component is of uniform strength and contributes equally to the overall strength of a device throughout its service life. Designing for uniform strength helps to achieve a design in which material capability is fully utilized throughout a component's volume while at the same time preventing "over-design" and avoiding stress concentrations and other "weak spots". This approach should be applied whenever economic circumstances allow.

Design for Short, Straight-Line Force Transmission. Direct and short force transmission paths generally help ensure minimum volume, weight, and deformation while at the same time allowing the use of simple component shapes that can be easily analyzed and manufactured. Application of this principle suggests that it is best, if possible, to solve a problem using tensile or compressive stresses alone, because these stresses, unlike bending and torsional stresses, utilize all the material and produce smaller and more uniform deformations. This principle is particularly applicable if a rigid component or support structure is needed.

Design Parts for Matched Deformation. When force is transferred from one component to another, stress concentrations often occur in the region where the components interface because individual stiffness, and therefore deformation of the components, differ. This often results in weak points, regions of high wear, and assemblies that are sensitive to manufacturing error (e.g., tolerance stack-up). Designing parts *in such a way that, under load, they will deform in the same sense and, if possible, by the same amount* helps to ensure smooth, gradual transfer of force from one component to another and helps avoid sharp bends and abrupt shifts in force flow direction.

Create "Self-Help" Designs. Functional enhancements and robustness can often be obtained through the appropriate choice of system elements and their arrangement to provide mutual supportive interaction. In self-helping design, an initial effect is combined with a supplementary effect to produce an overall effect. For example, by designing for Z-axis, top down assembly, gravity assists the insertion process. Similarly, a radial-shaft seal may seal initially due to elastic force developed by deformation of the seal around the shaft. As pressure increases, a supplementary pressure force acting on the seal is developed which increases the sealing force between the shaft and the seal. This type of design is *self-reinforcing* self-help because the initial and supplementary effects have the same sense and are therefore additive in their affect. A *self-balancing* type of self-help design is created when the initial and supplementary effects oppose each other. For example, by using opposite hand helical gears, axial thrust developed by the gear mesh places the shaft in compression, which induces beneficial compressive stresses that help offset (balance) the adverse effects of reversed bending caused by the radial gear forces and shaft rotation. The result is a design that is more resistant to fatigue failure. Shot peening and pre-setting of springs illustrate balancing effects that are induced during component manufacture. *Self-protecting* solutions form a third type of self-help design. In these designs, the supplementary effect arises due to a change in load path that is caused by an over-load such as when a helical compression spring is compressed solid.

Design for Stability. The design should be constantly checked for possible instabilities that could develop due to internally or externally induced disturbances. To avoid instabilities, the product should be designed in such a way that the disturbances either cancel out or are reduced as much as possible. The seal design concept shown in Figure 13.1 is a good example.

Optimize Design Parameters

Design parameters are design variables such as dimensions and spring rates over which the designer has control. Robustness of the design can be improved by specifying parameter values that reduce sensitivity to hard-to-control variation as illustrated in Fig. 13.2.

Robust Design: Combating Variation and Change

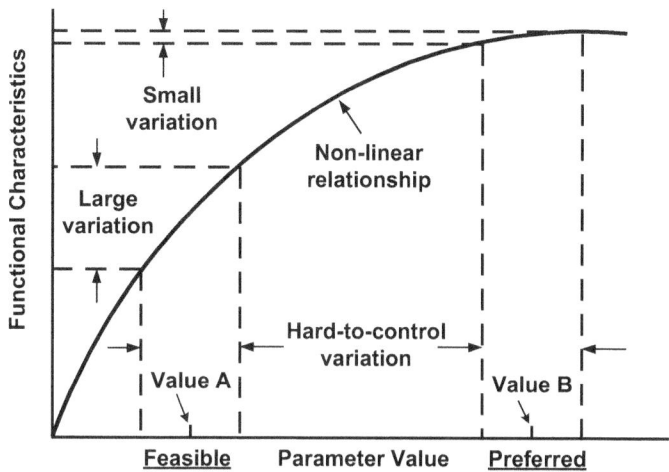

Figure 13.2 Value B minimizes variation in functional characteristics.

Illustrative Example 13.1: Select a helical coil spring that is to be used in an application where spring force should be as constant as possible, but the working length (deflection) of the spring is hard-to-control. Two springs are available: a relatively stiff Spring A, which has a high spring rate and a softer Spring B, which has a low spring rate. Which spring should the designer choose, A or B? Looking at Fig. 13.3, we see that the softer spring B is preferred because the force variation (ΔF_B) over the range of hard-to-control deflection variation is considerably less. Generally, when spring stiffness is unimportant, a softer, more flexible spring is preferred because it is less sensitive to hard-to-control variation in working length as well as variation of helical spring manufacturing parameters such as free-length and number of turns.

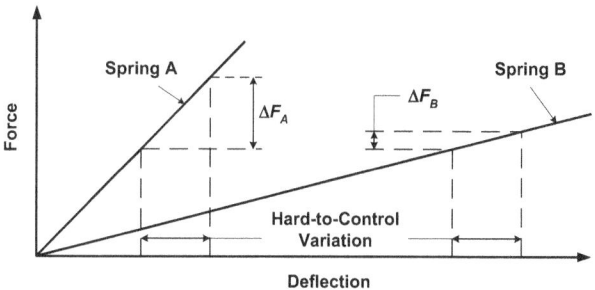

Figure 13.3 Spring-rate curves for spring A and spring B.

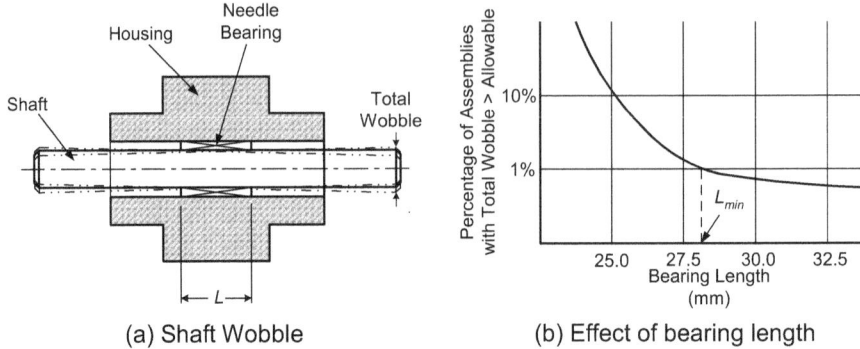

Figure 13.4 Bearing assembly: (a) unacceptable wobble and (b) effect of bearing length (L) determined from a "Monte Carlo" simulation of tolerance stack-up.

Illustrative Example 13.2: The bearing assembly shown in Fig. 13.4a wobbles unacceptably due to hard-to-control tolerance stack-up of the shaft, needle bearing, and housing dimensions. What minimum bearing length (L) is required to ensure that the percentage of assemblies exhibiting unacceptable wobble is below 1.0%? The percentage of assemblies that exceed the allowable wobble is predicted as a function of bearing length using commercially available "Monte Carlo" simulation software. As shown in Figure 13.4b, this percentage varies non-linearly with bearing length (L). It is seen from this curve that the percentage of assemblies that wobble unacceptably will be below 1.0% when L > 28-mm. This is a relatively simple example because only one design parameter (L) is involved. In situations involving more than one parameter, identifying robust parameter values becomes significantly more complicated. For these cases, design of experiments (DOE) techniques such as the Taguchi method (Taguchi, 1993) are recommended.

Design to Minimize Variation

Minimizing the source or cause of variation or error can be another effective means for implementing robust design. For example, by providing features that help orient parts and facilitate assembly, the opportunities for manufacturing error can be greatly reduced (see Fig. 13.5). In general, the goal in this approach is to design the product in such a way that it is easier to "do it right" than it is to "do it wrong".

Robust Design: Combating Variation and Change

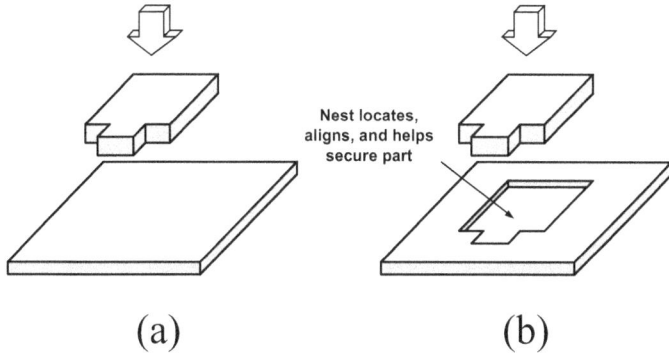

Figure 13.5 Example of design for reducing variation and error: (a) this part could be placed in any orientation, (b) this part has a nest that helps insure consistent orientation from product to product.

Design-In Variation Tolerance

Another approach to robust design is to provide features that are variation tolerant. For example, providing a feature that is easily recognized by a vision system, regardless of lighting conditions, makes the vision system less sensitive to hard-to-control variations in lighting conditions. Similarly, use of generous tapers and other guiding features makes part insertion less sensitive to placement accuracy of an assembly robot. Use of selective compliance is often an effective way to relax tolerance requirements (see Fig. 13.6). Replacement of a mechanical adjustment with a spring-loaded support that compensates for wear, not only eliminates the need for adjustment and service, but also helps make the product robust against deterioration over time.

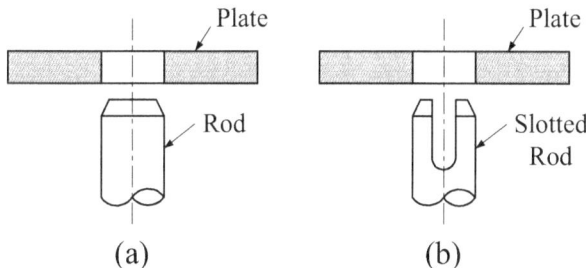

Figure 13.6 Example of "designed-in" variation tolerance: (a) two stiff components require tight diametrical tolerances, (b) selectively compliant member (slotted rod) allows much looser diametrical tolerance.

Summary of Key Concepts

- A robust design is one that is insensitive to and/or tolerant of change and variation in operating conditions, manufacturing conditions, and business conditions. True product robustness is "designed-in".

- Optimizing the geometric layout for maximum total design value is the secret to robust design.

- Design for change is the secret to long-term robustness of the manufacturing system as a whole.

Chapter 14
Coordinated Design: Integrating Product and Process

Coordinated design is a philosophy of design that seeks to harmonize and optimize the geometrical layout and proposed production concept together as an integrated whole. The geometric layout represents the "design" side of the equation. It assigns functional requirements to individual building blocks, defines the location and spatial arrangement of the building blocks, delineates the assembly structure that ties them together, and determines the sequence in which they are assembled. The proposed *production concept* represents the manufacturing side of the equation in which a mass production scheme to manufacture the design in the production quantities required is envisioned. The goal is to develop a conceptual understanding early in the project as to how the design is to:

1. Provide the required functionality.
2. Accommodate market and technological change, enable customization, and leverage the advantages of standardization.
3. Ensure ease of manufacture, assembly, and lifecycle support.

This chapter presents a six-step process for developing a coordinated product and process design concept. In the ideal, this process should be performed early in the conceptual phase because it helps sensitize the team to the opportunities available and creates a shared vision toward which to work. It is not unusual for the initial concept to change significantly as the design direction is firmed up and details of the design become more defined. The key is to use the coordinated design concept as a guide to ensure that "downstream" needs are constantly being considered. Because the process assumes a broad working knowledge of design and manufacturing, you might find it useful to reread the section on production lines in Chapter 3 and to review all of Chapter 4.

Geometric Layout

The geometrical layout spells out the way in which the goals of the design are to be achieved by the physical design. For designs that involve the assembly of parts, the geometric layout does this by defining the spatial arrangement and assembly of the parts so that the design is able to fulfil its intended function. As discussed previously in Chapter 4, the assembly structure together with the material and process selection define the geometrical layout of a design. In addition to these, it is often useful to impose an organizing principle in the form of a design architecture to help guide the choice of geometrical layout. *Design architecture* is the scheme by which the design is organized into physical building blocks or "chunks" as they have come to be called.

The design architecture is strategic in that it implements the design goals that are to be achieved by the design project. Design goals can range from improving functional performance to accommodating technological change and product variety. Often the goals will also include manufacturability considerations like part count reduction, eliminating product defects and rework, and simplifying testing procedures.

A *chunk* is a collection of parts, components, and sub-assemblies that are grouped together for a well-defined specific purpose. An individual part such as a large casting or housing can be a chunk. A standard off-the-shelf component such as a pump can be a chunk. A major sub-assembly can be a chunk. A group of parts such as gears in a speed reducer can form a chunk. At one extreme, the whole design might be lumped together as one chunk while at the other extreme, each individual part might be treated as a chunk.

Chunks are defined internally within the firm and need not have any meaning or importance to the end user or other external customers. A *module* is a chunk that is designed so that it is totally self-contained with standardized interfaces that allow it to be used interchangeably in different designs or within a family of products or design applications. Interchangeable camera lenses and designs that allow larger systems to be built up by using multiple smaller standardized repeat chunks (e.g., drawers used in a furniture cabinet) are examples.

Each chunk is composed of physical elements that implement the functions of the design. *Physical elements* consist of parts, components, and sub-assemblies that operate together to provide desired design functions. Some physical elements are dictated by the physical concept (e.g., the piston in an internal combustion engine), some by the function they provide (e.g., an electric motor), and some by the assembly concept (e.g., a frame or base component). A *standard component* is a physical element or chunk that has a pre-determined interface and is used interchangeably in a variety of different designs and applications. Standard "off-the-shelf" physical elements such as light bulbs, electrical connectors, and mechanical fasteners, are categorized as *external* standard components. *Internal* standard components, on the other hand, are standardized common parts, chunks, or modules that are unique to the firm and are used exclusively in its products.

Designed Components are unique physical elements such as a part, component, subassembly, or chunk that is designed to meet specified needs of a given design. Choosing between designed and standard physical elements can be critical. For example, the choice of designing a special electric motor, optimized for performance and weight, as opposed to purchasing a standard off-the-shelf motor, can have far reaching performance, cost, quality, timing, and manufacturability consequences.

Physical elements are designed to implement *functional requirements*, which are the individual operations and transformations that are required to provide the functionality of the design. As discussed in Chapter 4, functional requirements (FR's) are statements of the design problems to be solved. They are often distinguished as main, critical, or auxiliary FR's and are described verbally such as "enclose", "support", "channel", before they are reduced to specific technologies, components, or physical concepts.

Degree of Modularity

Design architectures are characterized by their *degree of modularity*. The two alternative radar antenna support designs developed in illustrative example 2.1 (page 15) are shown in Figure 14.1. Design A is an example of an integral architecture. *Integral architectures* exhibit one or more of the following characteristics (Ulrich, 2000):

- A single chunk implements several FR's.
- FR's are implemented using more than one chunk.
- Interaction between chunks are ill-defined and/or ambiguous.

A design that utilizes an integral architecture will frequently be designed for high performance. Many functional elements may either be distributed across several chunks or, what is more often the case, combined into several physical elements that comprise a single chunk. Integral architectures must therefore be carefully designed to ensure that FR's do not become coupled in undesirable ways. To further complicate matters, any modification to a physical component or feature may require extensive redesign of one or more chunks, or as in the case of Design A, the whole design.

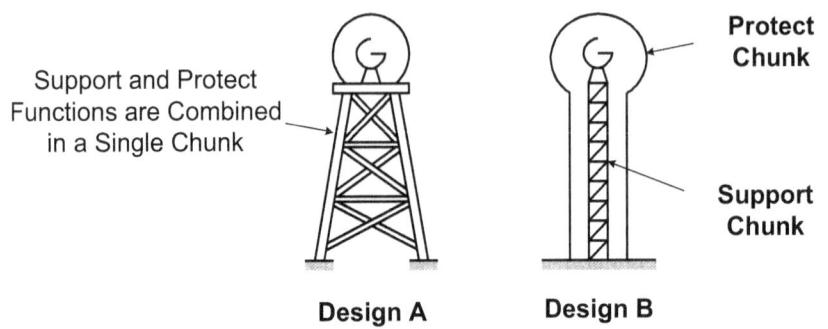

Figure 14.1 Alternative radar antenna support designs. FR's are (1) support the antenna, and (2) protect the antenna from external disturbances.

A modular architecture is the opposite of an integral architecture. As illustrated by Design B in Fig. 14.1, a modular architecture is one in which each functional requirement of the design is implemented by exactly one chunk. This means that a design change can be made to one chunk without requiring a change to any other chunk. It also means that each chunk can be designed independently of the others provided that the well-defined interactions between chunks is preserved.

Most practical design architectures fall somewhere in between these extremes. From a good design standpoint, architectures that exhibit a high degree of modularity are preferred because the independence of functional requirements (the independence axiom) is assured. Also, a high degree of modularity permits the detailed design and testing of individual chunks to be assigned to teams, individuals, and/or suppliers. This allows different portions of the design to be carried out simultaneously. This is what is so nice about the logical building block method discussed previously in illustrative example 4.3 (page 61). Not only does the method generate numerous alternative physical concepts, it also facilitates a high degree of modularity if each sub-solution is assigned to its own chunk and the interactions between each chunk is carefully controlled to maintain independence.

Chunking Approaches

The challenge in creating a viable design architecture is, of course, to find a logical approach for assigning the FR's to chunks. Depending on the design situation and the FR's involved, there are a variety of approaches. For purposes of illustration, a two-step process is suggested: (1) create a function decomposition schematic, and (2) logically assign elements of the design to specific chunks. This process is best explained by example.

Coordinated Design: Integrating Product and Process

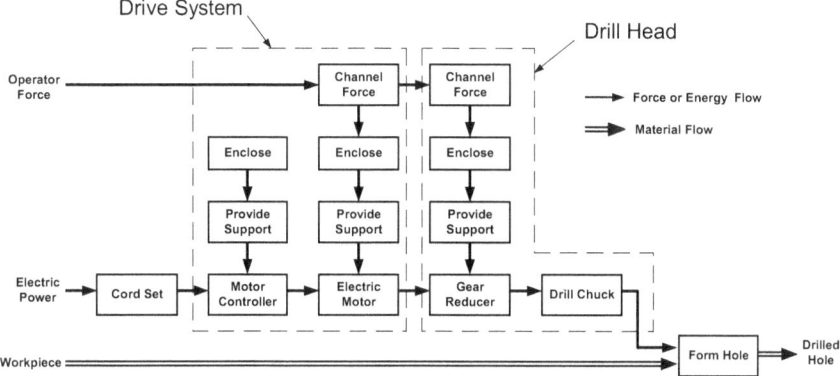

Figure 14.2 Function decomposition schematic for corded power drill.

Illustrative Example 14.1: The FR's of a corded power drill are to be assigned to chunks. Develop a possible functional decomposition schematic and suggest a possible alternative chunking approaches. Recall from Chapter 4 that a functional decomposition schematic is a diagram representing the constituent elements of the design. Some of the elements in the schematic are physical elements, some are defined physical components such as external or internal standard components, and some are functional elements that remain described only functionally. One possible version of a function decomposition schematic for the corded power drill is shown in Fig. 14.2. As illustrated, a potential chunking approach might be to create two major chunks: (1) the drill head, and (2) the drive system. Each of these chunks is created by grouping the appropriate elements of the diagram together as shown. The purpose of chunking in this way is to consolidate force-flow for high performance. An alternative approach is to assign each element to its own chunk and thereby keep the door open to innovation. This approach assumes that the right design architecture will emerge as the coordinated design progresses.

Illustrative Example 14.2: Disassembled views of corded power drill products sold by two different manufactures are shown in Fig. 14.3. Each product serves the same basic function, but the geometric layout and "chunking" of each differs. In Electric Drill A, the motor and housing (plastic injection molded base and cover) form a first chunk, the gear reducer forms a second chunk, and all remaining parts are each their own chunk. In Electric Drill B, the electric motor forms a chunk while all other physical elements are each their own chunk. Each layout offers advantages and

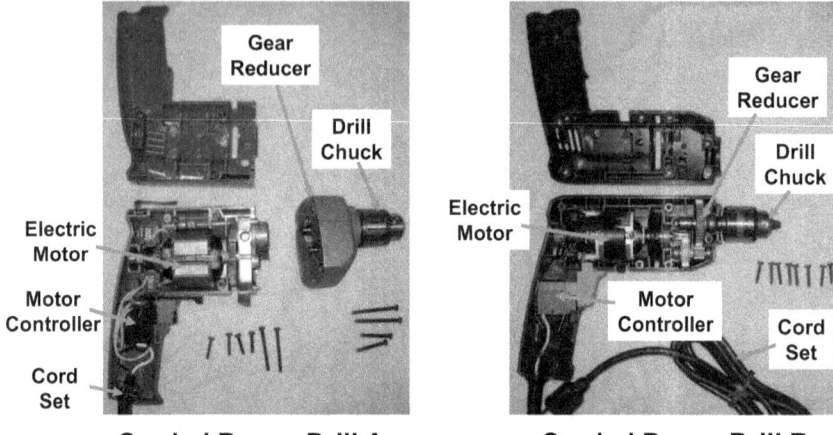

Figure 14.3 Alternative corded power drill geometrical layouts.

disadvantages. The geometrical layout of corded power drill A utilizes a drive system chunk and a drill head chunk like those identified in illustrative example 14.1. This allows the housing of the drill head chunk to be a metal die casting thereby greatly enhancing the heavy-duty performance of the drill. However, close examination also reveals many manufacturability issues. For example, the electric motor is integral to the drive system housing. As a result, manual assembly of the spring load brushes is most likely a necessity. In addition, a total of twelve screws involving five different types are used. Corded power drill B, on the other hand, has been designed for assembly. All parts are inserted top-down, the bottom half of the drill housing has molded-in fixture nests to locate and hold all parts until the cover is secured using six identical self-tapping screws to complete the assembly. The electric motor is a self-contained, standardized chunk making the option of automated assembly possible. In addition, most of the parts, other than the plastic injection molded housing halves, appear to be either internal standard PM parts (gears, etc.) or purchased parts (cord set, etc.). Both products appear to be using standardized motor control chunks.

As these examples illustrate, it is important to consider several different function decomposition schematics as well as different chunking approaches. The best geometric layout is seldom the first one that is envisioned. The best geometric layout usually gradually emerges as the conceptual design process progresses and more complete design information becomes available.

Production Concept

Each level of the design (part, sub-assembly, final assembly, etc.) requires a manufacturing plan (fabrication description). The *production concept* is the manufacturing plan for producing the finished design. If the finished design involves assembly, then the production concept will embody a detail assembly process plan and assembly method, together with information detailing how the designed components, standard components, and sub-assemblies are to be transported, handled, inserted, retained, secured, checked, and tested for defects and acceptable performance. When possible, it is highly desirable to develop the production concept concurrently with the geometrical layout to help ensure that production requirements and constraints are considered early. Undesirable interactions are avoided, information content is minimized, and potential problems are eliminated before they become real problems on the manufacturing floor.

The *assembly process plan* details each assembly step and operation required to complete the finished product. It is developed iteratively by dividing the assembly into rational work elements and considering the alternative assembly sequences possible given the precedence constraints established by the geometric layout and the assembly method to be used. Three basic *assembly methods* are available: (1) manual assembly performed at a bench or on a transfer-line; (2) special-purpose transfer machine assembly (fixed or flexible automation) in which assemblies are either transferred by an indexing transfer device (rotary or in-line) or by a free-transfer non-synchronous device; and (3) robotic assembly. In practice, the production concept will generally involve a combination of one or more of these methods.

Illustrative Example 14.3: Consider an automobile design in which the geometric layout consists of three major chunks: (1) the body-in-white (frame), (2) the rear axle assembly, and (3) the power train assembly. As shown in Fig. 14.4, a possible production concept for this geometric layout is to (1) load the body-in-white onto the assembly line, (2) assemble interior and exterior components, (3) add the rear axle chunk and the power train chunk at the final two assembly stations.

Production Concept Development

Development of a production concept begins with the system requirements (production quantity, effective time, etc.) and parameters

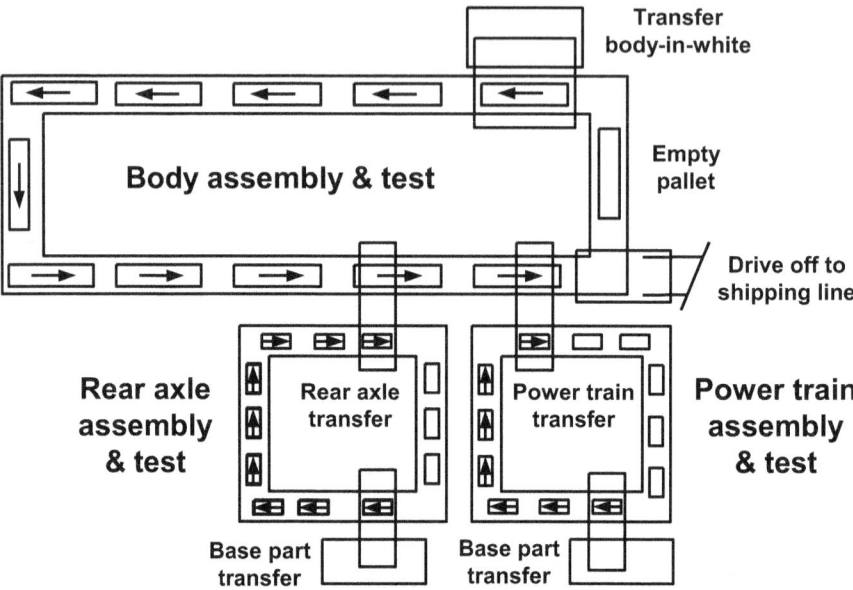

Figure 14.4 A possible production concept for illustrative example 14.3.

determined by the geometrical layout (work content, etc.). These requirements drive all subsequent decisions. Production concept development translates these requirements into a production line design that can produce the product at the rate required. The development process involves at least the following activities (Whitney, 2004):

- Choice of production resource (people, fixed automation, flexible automation),
- Development of a process operation sequence.
- Development of a process flow including the selection of the number of workstations and assignment of operations, work tasks, and processes to be performed at each workstation.
- Layout of the line including choice of system architecture (see Fig. 3.4, page 29), material handling methods, and location of part storage facilities.
- Layout of each workstation including choice of part handling method, location of part storage facilities, methods design (i.e., resource motions, activities, processes, etc.), and so forth.

Coordinated Design: Integrating Product and Process

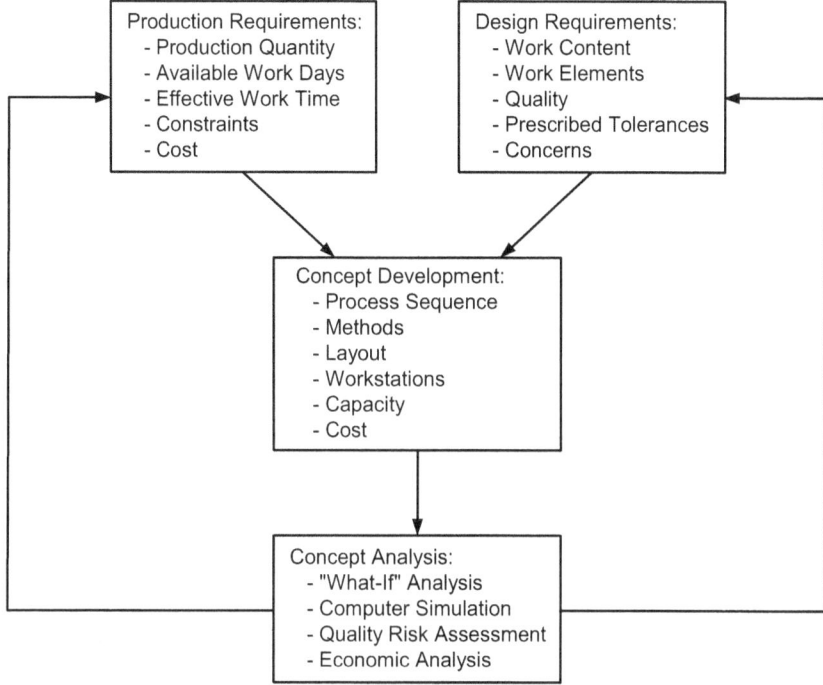

Figure 14.5 Production concept design-analyze-redesign process.

Once a preliminary production concept has been created, it is analyzed and improved through an iterative process of analysis and redesign. Typically, the analysis phase will include cost analysis, methods analysis, motion analysis, process failure analysis, and so forth. Based on results, the preliminary production concept may be modified and adjusted to correct deficiencies until an optimal design is achieved. Following this, some teams may create alternative production concepts that are then subjected to the above steps and their performance and cost are compared with the first one. This process may be repeated many times, depending on the design team's imagination and/or time available, until a satisfactory production concept has been determined. In some cases, the final concept may be a combination of the best aspects and features of two or more alternatives. The design-analyze-redesign process is illustrated in Fig. 14.5.

The development of workstation concepts may also be included as part of the production concept development. Each workstation on the line must be designed to perform the tasks it has been assigned. This involves consideration of a variety of factors including part presentation, tools and fixtures, transporting work units into and out of the station, displaying instructions, recording data, and generally making it possible to perform the assigned tasks as quickly as possible. The workstation resource must be able to reach everything, do the work efficiently and effectively, and, if it is a human operator, remain comfortable, confident, and safe. To achieve successful performance of the workstation, the workstation layout must ensure that all parts and tools be logically arranged and the transport and logistics of parts, tools, fixtures, and so forth must be carefully planned. Part feeding is another important consideration. Parts must be brought to the point where they will be used and, importantly, they must always be under control so that they stay intact and clean, and so that none get diverted or lost. Workstation material flow can have a profound effect on the cycle time of the workstation. The ideal situation is one in which the parts are ready, in the correct orientation, the moment the resource needs them. When the design is coordinated with the workstation concept, processing time is minimized, and no processing time is wasted.

Much of manufactured quality is achieved at the workstation level. Processing operations happen very fast, and operators can easily fall into repetitive activities in which they stop paying attention to what they are doing. It is therefore necessary to guard against mistakes by "mistake-proofing" the workstation. One of the most effective ways to mistake-proof is to design the parts so that they cannot be assembled incorrectly. This is another one of the tangible improvements that DFM brings to the production concept. For example, the team can design a part so that it "nests" in an asymmetrical recess molded into the mating part, thus making incorrect part placement impossible. There are also many things the workstation designer can do. For example, screwdrivers with overload clutches that prevent over tightening can be used. Lights can be placed over part bins to tell the operator which part to select. Operators can be shielded from distractions that cause them to lose concentration and then make a mistake. Sometimes the use of automated equipment can be justified based on the need for high quality because machines do not get tired, do not get distracted, and perform the job the same way each time, every time. This is often true for assembly jobs that require rote repetition. However, if judgment is required, even if it is only to quickly check that a part is suitable for assembly, the job may be impossible or uneconomical to mechanize. Sometimes an appropriate compromise is to assign operators to make kits of known good parts that are then assembled by machine.

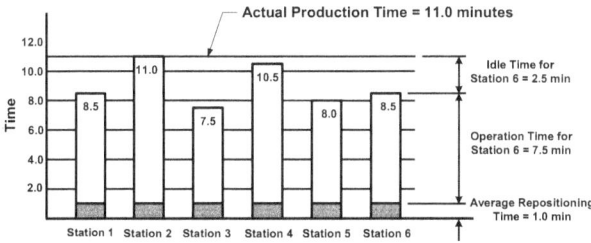

Figure 14.6 Actual workstation cycle times for the conventional approach in Example 14.4.

Development of a production concept can begin as soon as a viable candidate design concept is available. In the DFM approach, concurrent engineering is used to consider both the design and the production system simultaneously. In any case, it is vital not to wait until the product design is "done" because the most desirable results are obtained when the product and process are coordinated. In addition, when possible, the principle of "least commitment", should be followed. This principle requires that no irreversible decisions be made until necessary. Once an irreversible decision is made, both design and production options and choices become constrained. The principle of least commitment provides the time and conceptual maneuvering room needed to iteratively optimize the design and production concept as a coordinated system.

Illustrative Example 14.4: A company wishes to manufacture four versions of a product for a total of 15,000 product per year. Work content is estimated to be 45 minutes. Based on an effective work time per shift of 420 minutes and 2 shifts per day for 240 days/year, the production time is calculated to be 11.4 minutes/unit after being reduced by 15% to compensate for uncertainty. Develop an appropriate production system (a) using the conventional approach and (b) using a coordinated design approach.

(a) ***Conventional Approach***: The design is released to manufacturing. Manufacturing engineering analyzes the design and decides to use a sequentially arranged multiple workstation synchronous transfer line in which all operations are performed manually by human workers and the repositioning time is 1.0 minute. Although the theoretical number of workstations is 4, because of model mix, assembly sequence, precedence, and repositioning time constraints, an assembly transfer line having 6 workstations is required. Estimated cycle time for each workstation is shown in Fig. 14.6.

Table 14.1 Coordinated design savings of Example 14.4

	Conventional	Coordinated
Work Content per unit	45.0 min	24.0 min
Minimum number of workstations	4	2
Actual number of workstations	6	2
Actual production time per unit	11.0 min	12.0 min

(b) *Coordinated Design Approach*: The team designs the product to be assembled in its entirety by one worker at one workstation. Parts are delivered using kits that establish orientation and provide the required part mix for each product model. Information content of the geometrical layout and assembly sequence combination is minimized using the DFM guidelines of Chapter 9 through 11 to give a work content time of 24 minutes. This permits two parallel workstations, each having a cycle time of 12 minutes, to be used. In addition to the savings tabulated in Table 14.1, the coordinated design approach also reduces factory floor space, material handling complexity (and cost), and workstation cost.

Steps for Developing a Coordinated Design Concept

There are many approaches for implementing coordinated design. In general, most approaches will include the following six steps. Each step is discussed below in detail. The steps are:

1. Develop a comprehensive problem statement.
2. Create a function decomposition schematic.
3. Group elements into "chunks".
4. Visualize a geometric layout.
5. Envision a production concept.
6. Iteratively improve, refine, and optimize the concept.

Assuming that the coordinated design effort is performed early in the design process, the end result should, at a minimum, include descriptions of the major chunks, an approximate geometric layout, and an envisioned production line for manufacturing the design in the production quantities anticipated. The amount of detail developed during this six-step process depends on the design situation, how early the team is in the conceptual design process, and the level of effort involved. If the design is relatively uncomplicated, the whole exercise can be completed in a day or two long off-site brainstorming session. In more complex situations, the process may take place over extended periods of time as the physical concept and design direction is firmed up.

Step 1: Develop a Comprehensive Problem Statement

To ensure optimal chunking of the design, it is important to fully understand long-term market, technological, manufacturing, and lifecycle support needs. How will market needs and customer needs change over time? What aspects of the design will eventually need to be redesigned to accommodate changing technological capability? What upgrades, add-ons, and/or adaptations will be needed? What aspects of the design can or should be standardized? What aspects of multiple versions of the design are common? What aspects of multiple versions of the design are unique? What are the manufacturing, assembly, testing, and service issues and challenges? The problem statement is a clear and concise statement of what the coordinated design is to achieve with respect to "downstream" processes and long-term business needs.

When possible, it is often useful to formulate the problem statement as a series of design goals together with constraints that must be satisfied. In developing the problem statement, it is important to (1) thoroughly understand the performance, manufacturing, and lifecycle support issues and problems associated with previous and/or existing designs; (2) articulate a design strategy that supports the business and marketing strategies of the firm; and (3) understand the future to the fullest extent possible, especially with regard to technological change and market change. Remember that customers for the coordinated design include the manufacturing floor, system houses and other suppliers, service personnel, and others who are involved with "downstream" processes. In addition to interviews, careful observation of existing production processes and servicing procedures can also yield a great deal of insight. Because coordinated design is so comprehensive, it is extremely important that the time and effort required be spent "up-front" to get the problem statement right. If an important customer need is missed or inadvertently overlooked, much time and effort can be wasted. Most importantly, an early misstep can endanger the whole project.

Step 2: Create a Function Decomposition Schematic

The function decomposition schematic is the starting point for developing the coordinated design. The goal is to coordinate the geometric layout with a logical, well-reasoned "chunking" strategy and production line concept, while also avoiding undesirable interactions and minimizing information content. The schematic should facilitate this by representing the functional requirements of the design in a way that provides helpful information and guidance regarding spatial relationships, interactions (material flow, force and energy flow, information flow, etc.), and design constraints (e.g., required use of standard components). In most cases, there is a lot of latitude in how the schematic is drawn. For this reason, as illustrated by Examples 4.2 and 4.3, it is often useful to generate several alternatives to gain multiple perspectives on how the optimization might be achieved.

Step 3: Group Elements Into "Chunks"

Chunks are collections of physical elements that are grouped together to provide explicit advantages. The "chunking" logic that is used is critical to achieving a coordinated design that minimizes information content of the design. Begin by assuming that each element in the part decomposition schematic will be assigned to its own chunk; then, using the problem statement and the following guidelines, group elements to provide desired benefits and advantages. Always use the problem statement as a guide.

- **Close force-flow paths locally:** Elastic deformation caused by force-flow can be a major source of undesirable interactions (see Chapter 7). It also adds cost, weight, and complexity to the design. Grouping elements so that force-flow paths are closed locally minimizes and controls deformation and eliminates undesirable interactions.

- **Isolate precision:** Information content is reduced when elements requiring precise location or close geometric integration are part of the same chunk. Avoid tolerance stack-up by confining precise dimensions to a single part.

- **Share function:** When a single physical component can implement several functional elements of the design, these elements are best grouped together. <u>**WARNING**</u>: always be on the lookout for undesirable interactions.

- **Leverage vendor capability:** A trusted vendor may have specific capabilities. Group elements to take advantage of such capabilities.

- **Group similar technology:** Group functional elements that utilize the same technology. For example, grouping all functions that are likely to involve electronics might allow the possibility of a single circuit board.

- **Localize change:** Design for change (Chapter 13) by isolating elements that are likely to change into their own chunk or chunks.

Coordinated Design: Integrating Product and Process

- **Accommodate variety:** Group elements together in ways that make it possible to upgrade, add-on, and/or adapt the design according to changing customer and market needs. Isolate differentiating elements in one chunk.

- **Avoid manufacturing, assembly, testing, and service problems:** Group elements in ways that simplify and/or avoid production problems and assembly precedence constraints. Group elements to facilitate subsystem testing as well as quick maintenance and/or service.

Step 4: Visualize a Geometric Layout

This is the key step in the process. In addition to working out the dimensional and spatial relationships between the chunks, this key activity requires that an assembly structure (support and integrating scheme) and assembly sequence be envisioned. It may also be necessary to select a material class and/or manufacturing process. Developing the geometric layout and envisioned assembly design not only reveals whether the geometric interfaces among the chunks are feasible, but also identifies assembly constraints and requirements early in the design. Optimize using the ideas and concepts presented in Chapters 6 through 13.

Step 5: Envision a Production Concept

The envisioned production concept may include a target number of workstations (or workers), a material handling scheme, workstation concepts, factory floor layout, and other considerations that are important to the design such as testing and calibration. In general, the more detailed the production concept, the more useful it is in guiding design decisions. The degree of detail will depend largely on the completeness of the problem statement, the expertise of the team, and the maturity of the design. The minimum requirement for effective envisioning of the production concept is to have manufacturing engineering strongly represented on the team.

Usually the desired coordination naturally occurs as the result of timely input, guidance, feedback, and insight provided by the team members. A more proactive approach, which is sometime called "process-driven design", may also be used. In *process-driven design*, the production concept (preferred assembly sequence, target number of workstations, factory floor layout, etc.) is developed first and the product is designed accordingly. Process driven design requires a sophisticated manufacturing organization. For this reason, mature products such as automobiles and appliances are often best suited for this approach.

Step 6: Iteratively Refine and Optimize

Because the coordinated design steps are tightly coupled, iteration is usually necessary to achieve optimality. Iteration implements the "design-analyze-redesign" DFM approach discussed in Chapter 5. Keep the following in mind when refining and optimizing the coordinated design:

- Experimentation using simple models and mock-ups can answer questions, raise new questions, and provide understanding that cannot be gained in any other way.

- In general, the function decomposition schematic created in Step 2 will not be unique. Similarly, many alternative geometric layouts and production concepts are possible. Generate and evaluate several alternative designs to ensure that the best possible coordinated design solution is identified and chosen.

- Some chunks may be complex systems in their own right. Each of these chunks may have its own scheme by which it is divided into smaller chunks. The coordinated design approach is applicable at all levels of design.

- Coordinated design depends on all product and manufacturing information and knowledge being available. Industrial design, systems houses, equipment vendors, tool suppliers, and other outside stakeholders should always be involved in decisions that concern them.

- Coordinated design is a way of thinking about geometric layout optimization. It should not be viewed as a step-by-step procedure. Rather, it should be understood as providing the discipline needed to help ensure that all "downstream" needs and constraints are properly identified and considered early in the design process.

Window Air Conditioner Case History

A window air conditioner is to be redesigned to correct several design issues and to improve its manufacturability. An exploded view and geometric layout of the current design is shown in Fig. 14.7. Examination shows that (1) the assembly structure (sheet metal chassis) lacks clarity with respect to assembly, (2) there is no logic to the design, (3) piece-parts are optimized for sheet metal forming processes, but not for ease of assembly, and (4) there is no plan regarding standardization, model differentiation, functional testing, etc.

Coordinated Design: Integrating Product and Process 203

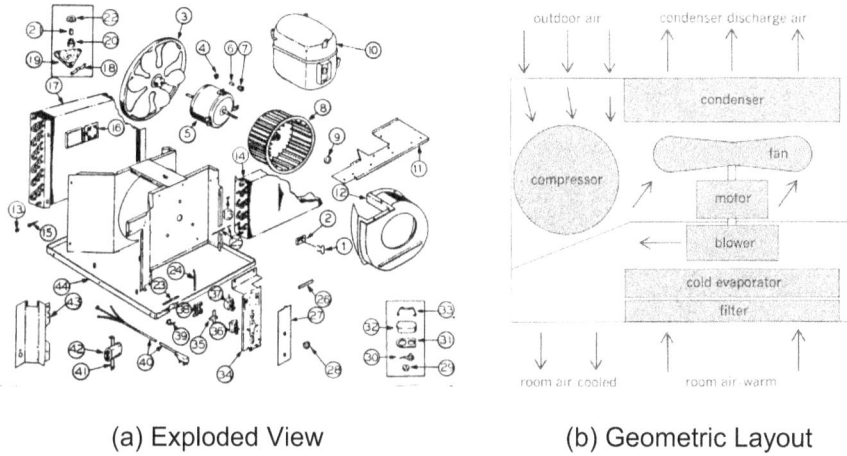

(a) Exploded View (b) Geometric Layout

Figure 14.7 Current window air conditioner design. (a) exploded view, (b) geometric layout.

The current window air conditioner design has been plagued with production problems. Primary among these are refrigerant leaks. As currently manufactured, leaking units are not detected until final testing of the completed assembly. Repair requires extensive disassembly, rework, and testing which is very costly and time consuming. Other manufacturing issues include noisy operation (vibration, buzz, rattle, etc.), marginal cooling performance due to air leaks and other quality problems, and unsatisfactory appearance due to poor fit and finish of the sheet metal components. Correcting these issues is the central concern of the redesign project.

Step 1: Develop a Comprehensive Problem Statement. The problem statement is formulated as four design goals.

1. Avoid adding value to defective units.
2. Reduce part count (currently 236 parts).
3. Simplify assembly and reduce work content.
4. Provide the ability to use different control technologies (mechanical, digital, etc.).

Constraints include the need to use the existing evaporator and condenser heat exchangers, which are internal standard components that are used interchangeably in several different products.

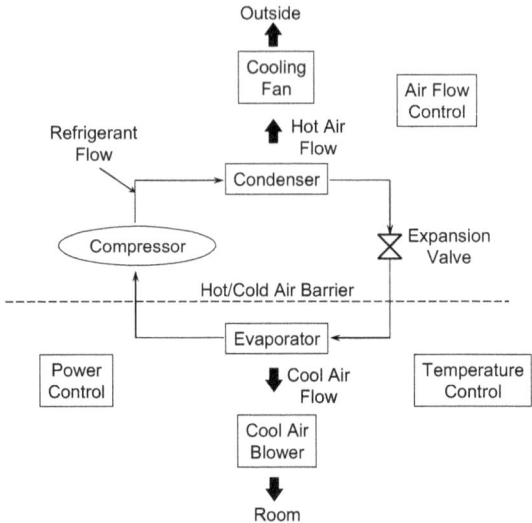

Figure 14.8 Window air conditioner function decomposition schematic.

Step 2: Create a Function Decomposition Schematic. A possible functional decomposition schematic is shown in Fig. 14.8. Since the air conditioner operates on the vapor compression thermodynamic cycle, the compressor, condenser, expansion valve, and evaporator are connected by copper tubing to form the refrigerant circuit. Also, a barrier is needed to separate the cold, room (evaporator) side of the air conditioner from the hot, heat rejection (condenser) side. Note that this diagram differs in various ways from that shown in Fig. 14.2. There are no hard and fast rules for creating a function decomposition schematic; the only requirement is that it help provide insight for developing a "chunking logic".

Step 3: Group the Elements into Chunks. Since refrigerant leaks are a major concern, group the compressor, condenser, expansion valve, and evaporator into a "refrigerant system" chunk. To isolate technological change, group the control functions into a "control module" chunk. Combine the fan motor, hot-air side fan, and cool-air side blower wheel into a "fan system" chunk.

Step 4: Visualize a Geometric Layout. Based on the DFM guidelines presented in Chapters 9 through 12, use a Z-axis stacked construction approach. As shown in Fig. 14.9, the coordinated design consists of a base and upper chassis, with the refrigerant system, control module, and fan system "sandwiched" in between.

Coordinated Design: Integrating Product and Process 205

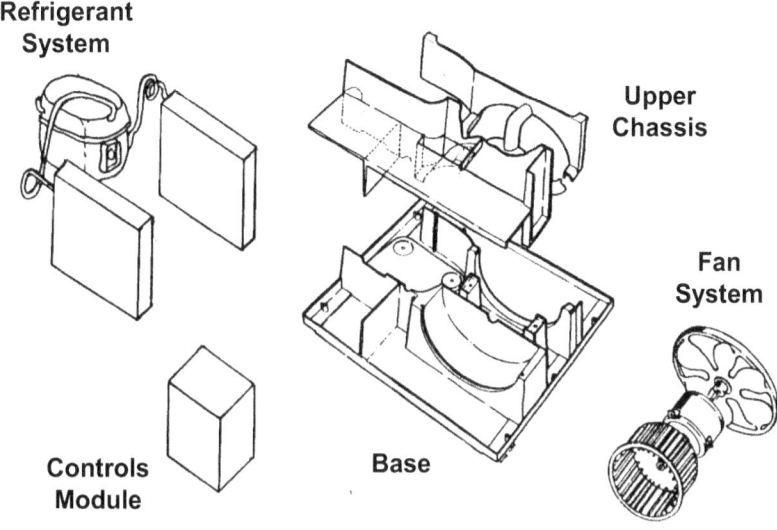

Figure 14.9 Visualized geometric layout and assembly structure.

Step 5: Envision a Possible Production Concept. In accordance with the design goals for leak detection and minimum possible number of assembly operators, an automated assembly concept consisting of a free-transfer assembly line with feeder lines is envisioned as shown in Figure 14.10. In this concept, each chunk is assembled and tested on its own feeder line and then robotically inserted into the final assembly.

Step 6: Refine and Optimize. To facilitate the chunking and stacked assembly concept, it is decided to fabricate the base and upper chassis out of sheet molding compound (SMC). This enables the integration of many sheet metal components into a single part. In addition, it makes it possible to mold-in nesting and guiding features as well as to form the hot/cold barrier and air-flow channels as integral features of the base and upper chassis parts. By eliminating air leaks, improved cooling performance using a smaller, less costly, and lighter weight fan motor becomes possible. Numerous other iterative improvements are also made as listed in Table 14.2.

Discussion of Insights Gained

By using the coordinated design approach to redesign the window air conditioner product, information content was greatly reduced, and several

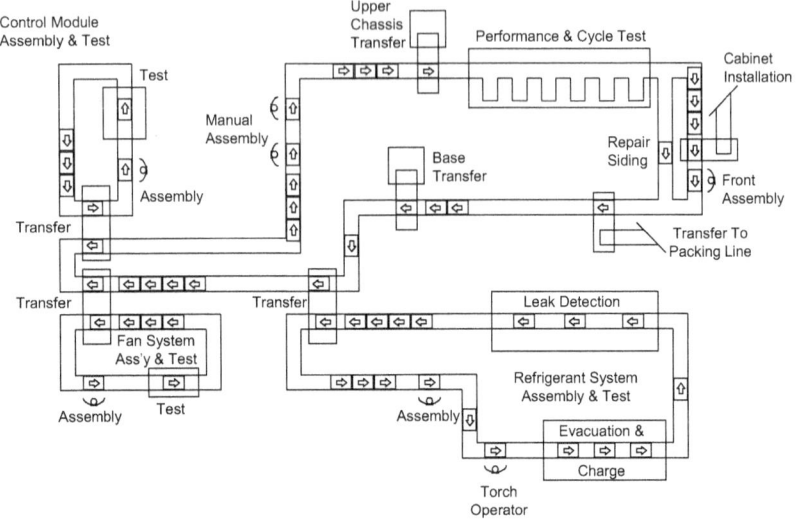

Figure 14.10 Envisioned coordinated design concept.

undesirable interactions were identified and eliminated. Insights gained from this redesign process are reviewed as follows.

- All design goals were met or exceeded. Experience has shown that setting "stretch" business, manufacturability, and lifecycle support goals as part of coordinated design stimulates innovation and design creativity. Most importantly, by considering these goals early in the design, they are almost always achieved, often in unique and unanticipated ways.

- Coordinated design challenges conventional thinking. In addition to improving air flow and cooling performance, the decision to mold the base and upper chassis rather than build them up out of numerous sheet metal components facilitated significant part count reduction and manufacturing simplification. This material substitution decision was not easy however, because it goes against many years of in-house sheet metal manufacturing practice and requires investment in new equipment and processes.

- Coordinated design facilitates early identification of undesirable interactions. Consider, for example, the robotic transfer and insertion of the refrigerant system that was originally envisioned (Fig. 14.10). Analysis shows that the concentrated weights of the compressor and heat exchangers together with the very flexible and

Table 14.2 Iterative Improvement of Window Air Conditioner Redesign

Current Design	Initial Redesign	Final Redesign
236 parts	155 parts	120 parts
	Issues: • Air handling • Refrigerant system installation • Too many fasteners, seals, etc. • Complicated tooling • Flexibility and strength of base component 26 performance concerns 43 assembly issues identified 9 tooling and process concerns	Major improvements: • Smaller fan motor • No fasteners in final assembly • All straight pull molds • Hardware commonality • Elimination of seals • Error proof assembly • Unique compressor mounting plate Process improvements: • Assembly sequence • Assembly line architecture • Coil and tubing connections relocated

"springy" copper tubing that connects them makes this operation prohibitively difficult, especially using robotic assembly. The problem was avoided by combining the refrigerant system feeder line with the final assembly line and using the base component as a fixture for assembling the refrigerant system. To facilitate this change, the tubing connections had to be relocated away from the SMC base. Doing this was easy, however, because the change was made early in the design. Such a change would probably not be possible after design release or after gaining UL approval.

- By using the DFM guidelines presented in Chapters 9 and 10, several seals and gaskets were eliminated. These design changes made it possible to eliminate two manual assembly stations from the final assembly line.
- As part of the design optimization, the base and upper chassis were redesigned to eliminate all camming and side-action in the mold. This not only reduced tooling cost and process cycle time; it also greatly simplified the air-flow channels. As a result, it was possible

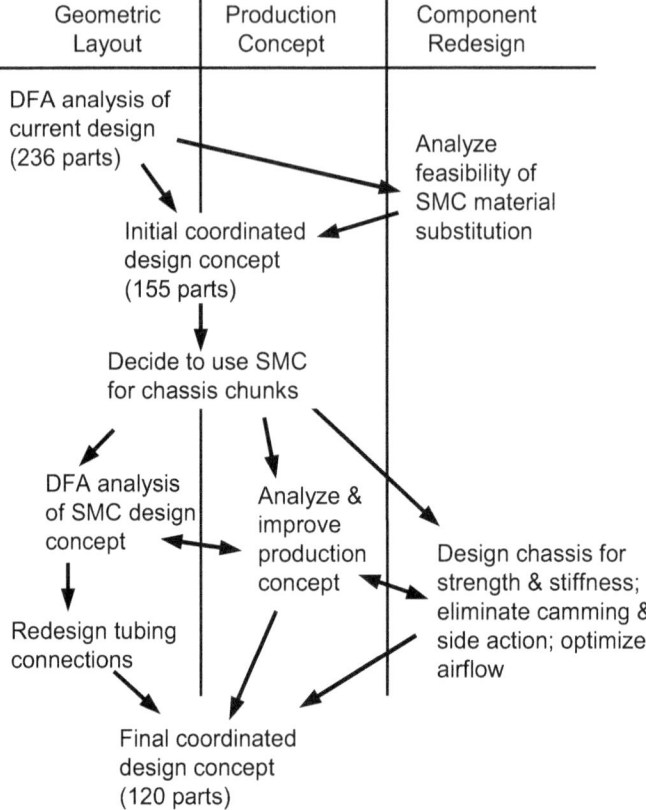

Figure 14.11 Coordinated design is iterative and non-linear.

to use a smaller and less costly fan motor, thereby producing a significant unanticipated cost and weight savings.

- The material cost for the SMC parts is greater than the cost of an equivalent spot-welded sheet metal fabrication. Because the firm had no experience with large molded plastic parts, making the tough decision to go forward with the SMC design required faith that the information content reduction would more than offset the increase in piece-part complexity. Quandaries such as this illustrate the challenges that can sometimes arise when seeking an optimal coordinated design.

- Although the six-step coordinated design process appears to be linear, the actual process of coordinated design is highly non-linear (see Fig. 14.11).

Summary of Key Concepts

- Coordinated design is achieved when "downstream" needs are considered early in the design using the DFM approach.
- Coordinated design is facilitated by grouping parts into "chunks" that provide well defined and understood advantages and then integrating the chunks into a harmonized whole by coordinating the geometric layout with an envisioned production concept.
- Coordinated design makes true global optimization of total design value possible.

Chapter 15
Ten DFM Success Factors

Implementing DFM can be a challenge even for large companies with established DFM programs. Efforts toward achieving "good" DFM seldom fully succeed the first time. For management and engineers alike, this can be frustrating, disheartening and defeating. One of the most important "lessons learned" about the hard job of "doing DFM" is that it requires constant nurturing and management commitment. Companies that succeed in implementing effective DFM generally do so over long periods of time. Just like everything else in manufacturing, DFM benefits from an attitude of, and commitment to, continuous improvement. In doing DFM, it is better to strive for excellence rather than perfection. Perfection is an impossible goal. Excellence, on the other hand, is an attitude. It allows for disappointment and prioritizes progress over "being perfect".

The 1955 Chevrolet: A "Fast Lane" Product Launch

In Chapter 1 of the book *Managing the Design-Manufacturing Process* (Ettlie and Stoll, 1990), a case history describing the design and development of the 1955 Chevrolet is presented. This was a car with a new body, a new platform (chassis and frame), and a V-8 engine (the first V-8 in a Chevy) that required the construction of a new, green field engine plant to produce it. It was launched from concept to pilot production in 24 months (summer of 1952 to summer of 1954) and dealers were fully stocked for customers by September of 1954 for the 1955 model year.

How was it possible for GM and Chevrolet to accomplish this? The engine and engine plant alone should have taken much more time than it did because the engine incorporated numerous unique innovations. For example, the block used only 9 major and 3 minor cores in the casting, while others of the day used as many as 22 cores to cast the V-8 block. Some of the most important findings and insights of the case include the following.

- The drawing board acted as the engineering conference room, exploiting group creativity. Design layouts and details were in plain view of everyone, making it quite easy to discuss and resolve design and manufacturing issues.
- All design for the car was done in-house.
- All engineering, including production engineering, reported to the same chief engineer.
- Tooling engineers worked with design engineers early in the design.
- Fewer people and fewer organizational levels were needed to make critical, innovative decisions or changes if they were needed.
- The focus of the project was to launch a new car, nothing more, nothing less. All new technology was proven, on-the-shelf, and ready to go. There were no hidden agenda items like, for example, learning about some new technology.
- Manufacturing engineers (called production engineers then) were charged with selling manufacturing people on the new design.
- Engineering changes could be initiated by either manufacturing or engineering, all were treated with equal importance, and all were resolved quickly.

These findings, together with ongoing experience with DFM implementation, suggest that successful DFM requires: (1) effective communication, (2) design discipline, and (3) a single-minded focus. Good communication makes needed information available, design discipline ensures a systematic approach, and a single-minded focus keeps DFM front and center. So, how can these qualities be instilled into the company culture and design process to help ensure DFM success? The following "DFM success factors" provide some useful insights.

1. Companywide DFM Mindset.
2. Focus on "Doing It Right the First Time"
3. The Principle of Least Commitment
4. Technology Readiness
5. DFM focused Design Reviews
6. Availability of Manufacturing Expertise
7. "Downstream" Input
8. In-house DFM training
9. Continuous Improvement of Product and Process
10. Identify and Fine Tune DFM Best Practices

Companywide DFM Mindset

Successful DFM begins with recognition that DFM is not the responsibility of design engineering alone. Rather, DFM is a companywide initiative that must be supported by a companywide attitude that values and supports DFM as a strategic activity. All levels of management, beginning with the CEO, as well as all employees should share the belief that DFM is essential to the success of the company. Often, this belief can be instilled by an overall directive that motivates all employees of the firm and puts in place a supportive organizational framework that values DFM. This overall directive is most effective when it is formulated as a set of "expectation standards" that convincingly communicate basic attitudes such as "excellence", "to be the best", or simply "quality". Expectation standards help generate the organizational will to make DFM a part of the company's core DNA. With a companywide DFM mindset powered by the right expectation standards, it becomes possible for the company to allocate the time and resources required and to develop the design discipline needed.

Focus on Doing It Right the First Time

The importance of design to the firm's bottom line is widely recognized. As a result, many manufacturing companies work hard to improve the way they do design. These efforts frequently focus on the company's design process by understanding the activities involved in performing design and rationalizing these activities into step-by-step procedures that follow a clearly defined process. Activities that are not needed or have marginal value are eliminated, while those that are deemed important are simplified and streamlined. In most cases, the emphasis is on shortening the design cycle, while also ensuring that the design is profitable. Seldom does a company set out to deliberately focus on "doing it right the first time" as a means for improving its design process. And yet, this is exactly what the DFM approach (see Chapter 5) aims to do. In fact, "doing it right the first time" is what DFM is all about.

In many ways, iterative problem solving in design resembles successive approximation algorithms commonly used to solve equations in mathematics. In finding the root of an equation using successive approximation, one first guesses a value for the root and then checks to see how closely it satisfies the equation. A new estimate of the root is then made based on information gained from the previous estimate. The process continues until a sufficiently close approximation to the root is obtained.

Ten DFM Success Factors 213

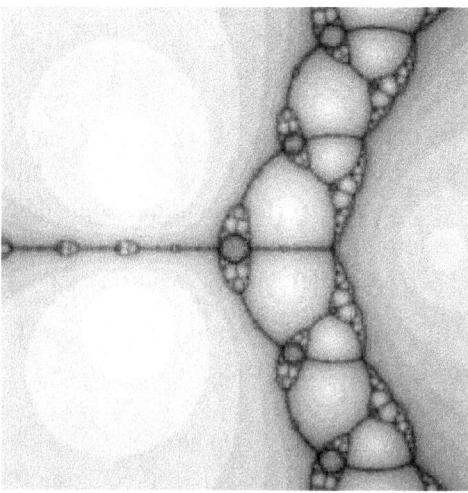

Figure 15.1 Newton fractal generated by the polynomial z^3-2z+2. Pixels are shaded according to the number of iterations it takes to get closer than 0.001 either to a root or to 0. Points in some basins never reach a root. (Uploaded to English Wikipedia by Henning Makholm, August 3, 2007.)

Over the years, mathematicians have learned much about this process. For example, the process can be made to converge more rapidly by using special methods, such as the Newton-Raphson method, to help select the next approximation or estimate. When more than one root exists, the initial value tried often determines the root to which the process converges. Hence, considerable time and effort can be saved by studying the equation carefully to select the best starting value. Mathematicians also know that if the initial guess is not carefully selected, the successive approximation process may degenerate into chaotic behavior where convergence becomes impossible. A graphic representation of the Newton-Raphson process as it is used to solve a third-order equation is depicted in Fig. 15.1.

When we look at the DFM approach (Chapter 5), we see a similar process. The problem formulation is the equation to be solved and the design solution represents the desired root. The design team studies the design problem and then creates an initial design which is then iteratively improved until an acceptable solution is obtained. Like the successive approximation process, the initial design determines what the final design is likely to be. If the problem of design is not well understood, then the design process may converge to the wrong design solution just as the successive approximation process may converge to the wrong root. And, like the use of the Newton-Raphson method in equation solving, the use of DFM design principles help accelerate the rate of convergence of the design process to the final solution. Finally, like a poorly chosen starting value for successive approximation, if the initial design concept is poorly conceived, it is possible that the process will converge to a marginal or inferior solution causing either the redesign process to fail or resulting in an inferior design being eventually accepted when the available budget of time and money is exhausted.

Although the successive approximation analogy is, at best, an oversimplification of a very complex process, it provides deep insight into the design process and how the process can be improved by focusing on "doing it right the first time". These include: (1) narrow the range of possible design solutions by imposing downstream needs and constraints on the design, (2) focus on getting the initial design right, and (3) use structured design methods to speed the convergence process.

DFM Narrows the Range of Possible Designs

If many different design solutions are possible, what is the one best solution and how can it be reliably identified? As we have seen throughout this book, DFM is the most effective and logical answer to this question. By superimposing "downstream" manufacturing, service, and lifecycle design needs onto the functional requirements of the design, DFM greatly reduces the number of plausible design solutions (see Fig. 15.2) allowing the best design to be identified quickly. And, because all downstream needs are included, the likelihood of "doing it right the first time" is dramatically increased. For companies that understand this, the motivation to make DFM a core component of their design DNA is a natural consequence.

Focus on the Initial Design

As in the successive approximation equation solving methods, the initial design can exert a tremendous effect on the final design that is ultimately evolved. If the wrong starting point is selected, then either a costly and suboptimal product is eventually designed, or the design program fails altogether. In general, it is very difficult to recover from a flawed starting point without literally beginning all over again and even then, there is no guarantee that a second try will fare any better. And yet, little time and money is usually budgeted for this most important first step.

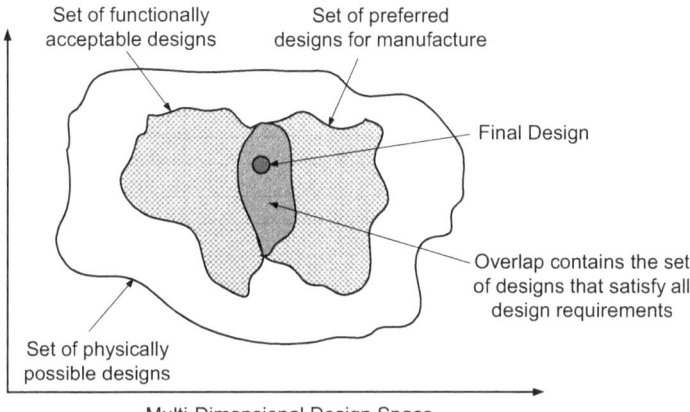

Figure 15.2 DFM narrows the feasible design region to a few best choices.

Most design teams are highly motivated to get started making progress right away. The feeling among management and the design team alike is that it is better to be doing something that looks like progress rather than to be making no apparent progress, which invariably looks bad to all concerned. Given this kind of mentality, sometimes a design team will pursue the first good idea they have without reflection or searching for alternative, and possibly better, design solutions. Often, the whole design effort is suffused with the belief that if bad design choices are made, they can be easily fixed later. The reality is that bad choices are seldom ever fixed, and the company is forced to live with the extra cost and customer dissatisfaction that results for the life of the product. In some cases, the product educates the marketplace on what to expect, forcing the company to live with its poor choices for many future product generations as well.

By focusing on "doing it right the first time", the company acknowledges the importance of the early stages of design. Instead of relying on marketing or some other project management entity to set project target dates based on an arbitrary timetable, this focus makes it logical and necessary to allocate time and resources that help ensure that the best possible initial design is identified and selected. The design process should not move on to the next phase until the design team is highly confident that the best initial design has been identified. More than anything else, this is the secret to doing it right the first time.

Use Structured Design Methods

In the Newton-Raphson method, the equation to be solved and its derivative are evaluated using the initial guess and the results are then used to generate a new, improved estimate of the root. The "Newton-Raphson" method illustrates how the DFM approach as well as "structured design methods" such as the geometric layout improvement method (Chapter 12), the Boothroyd-Dewhurst design for assembly (DFA) method, failure mode and effects analysis (FMEA), and computer simulations, use design information embodied in the initial design to suggest a new, improved version of the design. The actual work of generating the new version is, of course, performed by the design team or designer, but it is by using the structured design method that design insights are generated, creativity is stimulated, and ultimately, convergence to the best final design is accelerated.

Principle of Least Commitment

This DFM success factor is subtle but effective. It is aimed at short-circuiting the ripple effect and counter-acting the natural tendency to get off the dime and make recognizable progress, good or bad. Formally stated, the principle of least commitment requires that, in progressing from step to step in the realization of a design, *no irreversible decision should be made until it must be made*. This policy works to maintain maximum design flexibility in each step of the design process. Although postponing decisions seems counterproductive, it makes eminent sense because it keeps the door open for on-going innovation while also allowing for design change when unanticipated problems arise or are discovered.

Most importantly, the chances of "doing it right the first time" are greatly enhanced if irreversible design decisions are postponed for as long as possible. These are the design decisions that make design change difficult and costly. By postponing these decisions for a long as possible in the early conceptual phases of design, the design remains flexible and fluid, thus maintaining the maneuvering room needed to identify the best initial design. Effective implementation of this principle clearly requires belief and discipline on the part of both the design team and management.

I have personally witnessed the efficacy of the principle of least commitment on several occasions. In one project, the problem was to design a complex piece of manufacturing machinery to perform a difficult set of processing tasks at high production rates. The company chose to take a "skunk works" approach by challenging its "in-house" design organization as well as an external "design house" to both develop a viable conceptual design. The internal design group, who were very experienced in this type of machinery, immediately chose the approach they had used successfully in the past. This decision was in effect an irreversible design decision that severely constrained all that was done subsequently. The outside firm, on the other hand, seemed to spend its time asking questions and requesting seemingly unimportant and irrelevant information. Even late into the project, it seemed that they had not really come up with anything tangible and were just "wandering in the wilderness". When the deadline arrived, the internal design team was still wrestling with several problems and was not able to complete their concept proposal. The outside team, on the other hand, presented a brilliant concept that was totally original. By postponing irreversible decisions, the outside team kept the door to innovation open.

The principle of least commitment has been put into practice in a variety of ways. Flexible manufacturing systems (FMS) are an example. By utilizing programmable automation to quickly accommodate changes in product or production conditions, this approach to manufacturing makes it possible to postpone irreversible decisions indefinitely, long after the design has been released. "Chunking" (see Chapter 14) and other building-block standardization schemes (see Chapter 8) also offer many other opportunities for implementing this principle as an intrinsic feature of the design. When taken to its logical limits, the principle of least commitment can potentially form the innovative heart of a line of products, or possibly, as in the case of the Volkswagen MQB automotive platform design discussed in Chapter 8, an entire portfolio of products.

Technology Readiness

Nothing can short circuit DFM quicker than the need to develop new technology as part of a design project. For successful DFM, all technology must be proven, on the shelf, and ready to go before it is used in a new product or existing product development. This is because trying to solve technology problems on the fly as the product is being designed for manufacture inevitably guarantees that required design information will be incomplete. As a result, DFM will be difficult or impossible to achieve. Marketing and other entities within the firm may argue convincingly that a new technology be used in the design, but if avoiding costly and time consuming design change as well as sub-optimal design is a concern, it is almost always best to either not use the new technology or to postpone the project until the new technology is proven and ready to be used. Anything short of this sets the design team up for failure and spells disaster for customers and firm alike.

DFM Reviews

Because design is an open-ended process, there is no inherent self-correcting mechanism that acts to ensure that "downstream" manufacturing and lifecycle needs have been appropriately considered. *DFM reviews* are therefore an essential tool for ensuring that the right problems are being worked on and the right solutions are being selected and developed. Often, several DFM reviews may be useful during the course of design, especially for large projects where many subsystems must be integrated into a final design.

DFM reviews are systematic evaluations of the design from a manufacturability perspective. In general, DFM reviews should be scheduled by the team when milestones or important decision points have been reached. The design team (core team) must prepare for the review and they must have a clear idea of what must be demonstrated to pass the review. If problems or concerns are identified, then a detailed plan to correct the deficiencies should be developed and agreed upon as part of the review process. The plan should include a specific acceptability criterion that must be satisfied to move on as well as options to be considered if the problem or problems cannot be satisfactorily resolved.

When possible and appropriate, it is recommended that knowledgeable design and manufacturing personal who are not directly involved with the project conduct the DFM review. Asking the right questions about "why" and "how" can be critical to DFM success. The goal is to not only surface potential problems, but to also impose design discipline on the team. If the team knows from experience that hard questions will be asked and that they will need to justify their design decisions, then they are more likely to make more informed decisions that are easily defended. The result will be a more producible design. To ensure a successful review, the team should have an attitude that is open to constructive criticism and receptive to suggestions. To the extent possible, they should try to understand all criticisms and should seek creative approaches for how they might be addressed.

The DFM review should cover all aspects of the design including design alternatives that were or are being considered, cost comparisons between design alternatives, and supporting CAE analysis, computer simulations, and other relevant design issues, concerns, and deliberations. Manufacturing considerations are as important as the design. It is vital that, in addition to the design concepts, manufacturing concepts be reviewed and discussed in terms of design requirements and process constraints. Throughout the review, emphasis should be placed on the geometric layout, estimated cost, production plan, and on the appropriateness of part configurations, material selection, and manufacturing process selection.

Manufacturing Process Expertise

The design of components such as castings and powder metallurgy (PM) parts for low-cost and production friendliness requires deep manufacturing expertise as do hard-to-control processes like welding and adhesive joining. As discussed in Chapter 11, having process knowledge available at all stages of the component design is essential for successful DFM. Companies that successfully leverage manufacturing process expertise generally pursue a two-pronged strategy. For processes that are only occasionally used, the firm develops a portfolio of reliable high-quality preferred suppliers and system houses with whom they partner. For key in-house manufacturing processes, they commit to developing and maintaining internal process expertise. This maximizes competitive advantage while also avoiding unnecessary cost and duplication of effort.

To illustrate the importance of in-house technical expertise, consider a company that manufactures hydraulic actuators as part of its business. This company has used arc welding to join the piston to the piston rod as it's process of choice for many years and, as a result, has developed deep in-house expertise in this process. Arc welding is undesirable, however, because it requires extra machining and surface preparation steps as part of both the piston and the rod manufacture as well as additional inspection processes to ensure weld quality and conformance to design specifications. To reduce information content of the design and the manufacturing process, the company decided to deviate from its long-standing policy and purchase a new automated friction welding process that avoided edge preparation and other machining complexities while also automating the quality inspection process. About six-months after placing the new automated joining process into production, weld failures started to occur world-wide. In pursuing a fix, the company discovered that the problem was due to a lack of deep "in-house" technological understanding of the new joining process and how to correctly control and inspect weld integrity to meet the needs of this particularly harsh application. Instead of saving money, the joining method change ended up costing untold amounts in field repairs, warrantee claims, and damaged reputation. The lesson is clear: always support complex, hard-to-control in-house manufacturing processes with appropriate technical expertise.

"Downstream" Input

This DFM success factor is based on the reality that, no matter how experienced or knowledgeable the engineer or design team, no one person or team can be an expert on everything. Therefore, input from "downstream" knowledge sources such as assembly operators, component suppliers, automation systems houses, tooling suppliers, "lead users", product service providers, and others is highly desirable. At the same time, because many of these people are not accustomed to making design decisions, it is often more productive to ask their opinion about a new design rather than asking them to come up with redesign ideas. Discussing a proposed new design serves as a conversation starter that stimulates the person to begin to volunteer ideas, all of which should be listened to and then evaluated later.

In the case of manufacturing process, tooling, and production machinery experts, however, the opposite is frequently true. In these cases, it is most desirable to have the expert be a contributing member of the design team. As discussed in Chapter 11, the insight and expertise that these experts bring to the design is invaluable. Obtaining "outside" input can be challenging, especially if suppliers or vendors are involved and the purchasing process involves vendor selection using a bidding process. Innovative companies that are doing DFM successfully, sometimes circumvent this problem by forming partnerships with preferred suppliers. Such arrangements allow the outside supplier to be directly involved in the client's design process.

In-House DFM Training

DFM training is a proven success factor. The key is to make sure the training gets used. For most firms, this requires that some type of mechanism be put in place to allow for the new method or technique to be applied. One approach that seems to work well is to do DFM training in-house and, when possible, use the training to kick-off the start of a new design project. The goal is three-fold: (1) teach the principles of DFM, (2) apply methodologies such as the Geometric Layout Improvement Method (see Chapter 12) to one or more current production products, and (3) use the ideas generated to seed the new project. Key elements of this approach include:

- **Set Goals and Expectations:** To ensure that the training effectively furthers the company's DFM goals, it is important that the person conducting the training meet with key personnel, tour facilities, understand specific company aims and opportunities, and define specific learning objectives (e.g., specific techniques, culture change, etc.). In this way, the training is coordinated with the needs of the company.

- **Training Site Location:** One of the challenges of in-house training is to ensure full attention by all participants. If participant offices, desks, and phones are nearby, the temptation to "sneak out" to check pressing business will be strong. To avoid this, an off-site training location should be considered. Also, at the outset of the training, the importance of being fully involved should be emphasized. By setting the proper expectation standard, peer pressure will often help encourage full involvement on the part of all participants.

- **Select Training Participants:** If the training is being used to kick-off a new project, all personal involved in the project including design, manufacturing, and quality engineers, as well as key marketing, purchasing, and management personnel should be included. In situations where the training is not tied to a specific design project, all those in design, manufacturing, marketing, and purchasing who are directly involved with new product design and development should be included. It is extremely important that all those in key roles be involved so that the DFM training can become a "common language" that everyone in the company speaks and understands.

- **Focus on DFM Principles and Practices:** It is important that the training focus on basic DFM principles and practices such as those discussed in this book. With all key players present in the same room hearing the same message, the training session is an opportune time to ensure that all company personnel become familiar with these basic DFM concepts. This will help DFM to become a common language that is shared and used by all disciplines throughout the company to communicate and pursue "design excellence."

- **Analyze Current Production Products:** Nothing is more eye opening than discovering the DFM improvement opportunities that exist in current company products. To facilitate this discovery, use the Geometric Layout Improvement Method (Chapter 12) to analyze one or more current production products in "hands-on" class projects. The "hands-on" projects are performed by first dividing the class into multi-disciplinary teams of no more than four or five and then challenging each team to analyze a product or subassembly and propose an improved redesign. After each team has presented their redesign, have the class develop a new redesign by combining the best features and aspects of the team redesigns.

- **Follow Through with Expectation:** This is a pivotal element of the DFM training. The training experience will quickly fade if not reinforced with proper expectation. This is the reason why it is often ideal to tie the training to a specific new or current design project.

- **Identify and Train DFM Champions:** Identify, and further train selected design and manufacturing personnel to be DFM champions within the company. Successful DFM requires an internal "sales force" of champions who continually encourage, nurture, and support DFM practice within the firm. DFM champions who are so inclined should be encouraged to become "internal DFM trainers".

- **Train Frequently:** DFM seems to be one of those concepts that can quickly take a back seat when people get busy and time pressures mount. Experience has shown that frequent refresher sessions help keep DFM front and center.

Continuous Improvement of Product and Process

Even with a concerted emphasis on DFM, the best design for mass production is seldom achieved the first time because design knowledge is not often complete. Continuous improvement is essentially a small step-by-step incremental improvement strategy. By pursuing a policy of continuous improvement, the firm facilitates on-going DFM improvement to both the product and process. In the continuous improvement approach, company employees involved in all aspects of the manufacturing system are encouraged to continually seek ways for improving their own performance as well as the product and process. This not only leads to good improvement ideas, but also helps employees to take ownership of their work thereby improving worker motivation. Since many of the improvement ideas that are generated come from "downstream" customers such as assembly operators and service technicians, they are usually easy to implement and typically do not require large investments, redesign projects, or radical process changes. The focus of continuous improvement is on efficiency and evolution. Over time, it results in the identification, reduction, and elimination of suboptimal aspects of the product and process in incremental, continuous steps rather than in giant leaps.

Identify and Fine-Tune DFM Best Practices

A *best practice* is a commercial or professional procedure that is accepted or prescribed as being correct or most effective. Firm's that seek out and use DFM best practices generally do DFM more effectively and consistently. Many of the DFM success factors discussed in this chapter are, in fact, DFM best practices. The key is to identify these success factors as being essential DFM best practices and then continuously improving them to better meet the needs of the organization.

Some examples of DFM best practices that I have promoted and encouraged over the years include the following:

Observe the production-consumption cycle: To gain knowledge about the use environment and production system, study it. Ask questions, interview users and workers, send out surveys, etc. To gain wisdom and insight, observe. When you observe, you learn what is really going on.

Always choose from multiple alternatives: The probability of making high quality design decisions increases greatly as the number of alternatives considered increases. This is because generating alternatives forces the team to explore the whole design space. Always resist the tendency to go with the first idea that you think of. Good DFM results from doing the hard work of generating and selecting from many viable alternatives.

Understand how components deform: In mechanical systems, experience has shown that the underlying cause of almost all hard-to-fix or hard-to-explain manufacturing and/or operational problems involves elastic deformation of components and/or structures. Remember that real materials deform under load. If force and force-flow occur at any stage in the production-consumption cycle, there will be deformation. To avoid undesirable interactions, understand how this deformation affects the design.

Challenge separate parts: More than anything else, separate parts drive total cost. Never blindly accept the need for a separate part.

Plan the assembly sequence early: Planning the assembly sequence forces the team to think about downstream processes. Features that facilitate assembly can be provided, potential production problems can be anticipated and avoided, and the transition into production can be performed more easily and more quickly.

Plan the wiring layout early: In designs that involve electrical interconnection, planning the wiring layout early forces the team to consider integration issues early in the design when they are easy to deal with. Wire length can be minimized, connector location can be optimized, and potential quality risks and production problems can be identified and avoided.

Analyze the design for improvement opportunities: No matter how carefully you have designed for manufacture and assembly, always analyze the design using the Geometric Layout Improvement Method (Chapter 12) at least once before release. Almost all designs can be improved using this method.

Standardize and rationalize everything: Limiting choices saves design time, leverages experience, and reduces total cost. Purchased parts are a good place to start, but all aspects of the design that involve multiple choices are targets. Remember that benefits accrue over time, so staying the course is essential.

Summary of Key Concepts

- ➢ For DFM to fulfill its promise, it must be nurtured and given long-term management support. DFM demands nothing less than a philosophical shift to an organization-wide focus on DFM.

- ➢ Making DFM an inherent part of the firm's corporate DNA, however, is not an easy task. Because each manufacturing enterprise is different, each has different priorities and cultural constraints. For this reason, every firm must find its own path.

- ➢ Fortunately, there are plenty of success factors and DFM best practices that can help. The key is to understand why and how they work and then, to continuously improve them to better serve the needs of the design and manufacturing organization involved.

Chapter 16
Producibility Checklist

Producibility is defined as "the ease of manufacturing an item (or group of items) in large enough quantities". It depends on the characteristics and design features of the item that enable its economical fabrication, assembly, and inspection and testing by using existing or available technology. Under this definition, producibility is recognized as being the desired resultant of DFM (see Fig. 16.1). A producibility checklist, therefore, is a template that lays out the characteristics and features that a producible design should possess. Such a checklist serves many uses. For example, it can serve as a guide for designing producible products and equipment. Similarly, it may be used as a gauge in DFM reviews to evaluate proposed designs.

The producibility checklist presented in this chapter is formulated as a list of design directives (i.e., guidelines). The checklist encapsulates and re-emphasizes many of the points that have been made in this book. Although by no means complete, it is representative of some of the best "directionally correct" advice available for "teaching" and "testing" good DFM. Correct use and adherence to this checklist will almost certainly lead to better performing, higher quality, more adaptable product designs that can be made in quantity for less cost and in a shorter time. The checklist is stated as generically as possible and is recommended for use in all design and development programs regardless of type of production planned (i.e., prototype, one of a kind variants, batch, or mass), degree of novelty (i.e., completely new, improvement of an existing design, or adaptive design), or type of industry (e.g., aerospace, machine tool, automotive, consumer products, etc.).

Universal Themes

Woven through the producibility checklist are universal themes that characterize a good DFM program. These include:

Figure 16.1 Producibility is the resultant of DFM.

- **Integration** of product, production, operation, and support system design using the concepts of concurrent engineering and the team approach.
- Adoption and development of system wide **standards** that facilitate desirable producibility characteristics such as interchangeability, interoperability, simplified interfaces, effective consolidation of parts and function, availability of components, and so forth.
- **Modularity** of product, systems, production equipment, and process plans, software, training, etc.
- **Robustness** of product, process and associated organizational systems with respect to external and internal disturbance factors.
- **Standardization and rationalization** of all aspects of the product and lifecycle processes. Standardization involves the reduction of numerous options to the most popular ones based on examination of currently existing designs; rationalization is the identification of the fewest number of options to be used in future designs.
- Effective use of **training** to develop and sustain skills needed for product and lifecycle processes.
- **Reliability**, **Availability**, **Maintainability**, and **Durability** (RAM-D) considerations are vital producibility concerns. For example, all critical materials, supplied parts, "off-the-shelf" standard parts, and "outside" services should be obtainable from multiple sources, with minimal dependency on less reliable sources when possible.

When successfully implemented, these characteristics combine synergistically to greatly enhance producibility. If the checklist is used as a design guide, this synergy and the producibility benefits it produces are maximized. If used as a design review checklist, unexploited opportunities to improve these characteristic or possibly to leverage this synergy may be discovered.

Producibility Checklist

The producibility checklist is framed in the form of design directives, guidelines, and/or design suggestions. The checklist is relatively independent of any stated method of manufacture or assembly. When correctly applied and adhered to, the checklist will result in a product that is inherently easier to manufacture, assemble, test, service, and maintain regardless of the extent or degree of manufacturing mechanization employed. The guidelines are based on three basic producibility principles: eliminate, simplify, and standardize where practical.

- ☑ **Design for a Minimum Number of Parts.** Always seek the minimum number of simply shaped components. Check all parts, both jointly and individually, for function. Eliminate redundant parts. Eliminate extra parts by integrating their function into multifunctional parts. For example, combine mating or contacting parts that do not move independently of each other into an integral composite part by considering alternative materials, net shape manufacturing processes, and so forth. When possible, avoid designs that require separate fasteners. If separate fasteners cannot be avoided, minimize the number and variety of separate fasteners that are used.

- ☑ **Develop a Modular Design.** Create desired functional variants from basic and variation modules. Satisfy common functional requirements with standardized modules (or "chunks"), allowing increased production of identical parts. Isolate technologies or functions that are likely to change in a separate module (adaptive module). Design assembly modules, allowing easy assembly; maintenance modules, allowing easy maintenance; recycling modules, allowing easy replenishment. Standardize module interfaces (e.g., mechanical, electrical, etc.) to maximize interchangeability and minimize "special or unique" requirements for any given module.

- ☑ **Design for Ease of Manufacture.** Eliminate adjustments where possible. Design to avoid randomness. Provide cable runs, eliminate dangling connectors and unrestrained parts. Provide self-locating features on mating parts to eliminate ambiguity. Provide subassemblies and modules with enough strength to allow handling without the need for temporary shop aids. Avoid part to part or part to process dependencies such as "related features", "fit at assembly", "matched pairs", and "true position". Standardize and rationalize the fabrication and assembly operations to minimize special tooling and a large inventory of tooling. Design parts to minimize the number of setups and operations, tooling changes, special tooling, process motion, reorientation, etc. Eliminate secondary operations; primary operations should not produce a secondary operation. Use a rationalized family or catalog of processing features and capabilities to define detail part geometry. Minimize diversity of materials and material specifications; select from a rationalized list of "fabrication friendly" materials that are readily available from multiple suppliers. Minimize part variations; select from a rationalized list of standard parts (fasteners, etc.) that are readily available from multiple suppliers. Create a standardized and rationalized catalog of part families. Use these "building block" parts where possible in all new designs. Design parts for smooth production flow: avoid "specials" or "designer signatures". Design parts for orthogonal and/or parallel fabrication since most fabrication equipment is inherently designed using orthogonal and /or parallel constructs. This can minimize programming and debug time.

☑ **Design for Ease of Component Fabrication.** Know the fabrication process and design the component to insure conformance. Design parts to minimize material waste, number of setups and operations, and cycle time. Provide simple, readily usable part features to facilitate component locating, fixturing, and handling during fabrication. Design parts to minimize the variety and number of tools and fixtures required. Design parts to minimize the variety and amount of inspection and process monitoring required. When possible, provide features that allow use of a set of standardized and rationalized tools, fixtures, and inspection methods.

☑ **Design for Ease of Assembly.** Develop a rigid base component, fitted with guide surfaces, nesting features, and securing features, on which to build the assembly. Provide features (grooves, holes, flats, etc.) on each component to facilitate handling and insertion during assembly. Minimize the number of assembly directions. Top down (Z-axis) assembly is best; use gravity, its free. Design for a minimum number of assembly reorientations. No reorientation is best. Keep assembly motions simple; avoid multi-motion assembly. Design parts to be stable after placement or insertion. Provide generous tapers, chamfers, radii, leads and other guiding surfaces to facilitate ease of component insertion. When possible design parts so there is minimal resistance to insertion. Design parts for easy access during placement and insertion, Avoid blind assembly operations and blind final assembly inspection.

☑ **Design for Part Handling and Presentation.** Design parts to make position easy to achieve and maintain during handling, Symmetrical components are best. If a part must be asymmetrical, then overemphasize the asymmetry. Design to avoid handling difficulties such as nesting, tangling, or sticking together. Whenever possible, avoid flexible components. Link parts together as part of web or package in a tube or magazine to preserve and maintain orientation.

☑ **Develop a Reliable and Durable Design.** Avoid design alternatives that require specialized or unusual operator skills or training. Avoid cluttered, chaotic, or confused installations of wires, hydraulic lines, and other interconnecting components. Avoid design alternatives that are sensitive to uncertainties such as weld strength, bolt preload, lubrication conditions, or voltage fluctuation. When possible, design assemblies so that components are in a low energy state so that external work must be done to produce a change in the system. For example, when assembled, snap fittings are in a low energy state whereas a preloaded fastener is in a high energy state. Use components and devices for which well documented information on failure rates and derating is available. Consider standardization and rationalization of life parts (e.g., ball bearings, switches, etc.). Design to avoid or minimize the number of wear surfaces and rubbing parts. Avoid the need for adjustments and calibration during manufacture and with use and over time. Avoid designs that are sensitive to hard-to-control processing variables (e.g., curing time and temperature, holding pressure, cooling rate, etc.) or to voids, inclusions, grain structures, or other hard to control processing consequences.

☑ **Develop an Easy to Service and Maintain Design.** Avoid design alternatives that require special or unusual skills and training to perform repairs of structural, mechanical, and electrical components. Avoid design alternatives whose repair involves lengthy process time (such as cure and setup in bonding, sealing, or fitting) to achieve strength, alignment, shape, and so forth. Design so that limited life components such as fuses and filters, as well as fluid couplings, seals, and other parts that are at above average risk of failure are visible for inspection and are accessible for easy scheduled maintenance, removal and re-installation. Provide enough hand and tool manipulation clearance for easy maintenance, adjustment, measurement without removal of interfering components. Avoid the need for special tools. Consider use of a rationalized set of tools. Ensure that all major components are easily identifiable by serial number or part number without removal or reorientation, Use a rationalized set of common connectors, hardware, mounts, threads, fasteners, lubricants, tubes, hoses, wires, and so forth to achieve maximum interchangeability and part availability.

☑ **Coordinate Product and Workstation Design.** Design so that each workstation has self-contained quality with respect to monitoring, adaptation, control, and reporting. Quality should be designed into the coordinated product and workstation design, not added as an afterthought. Strive for zero set-up time by designing so that quick change dies, fixtures, tooling, and feeders can be used. Consider using identical tooling for automated and manual workstations to reduce inventory size and associated tool management problems. Management of tool and fixture quality (wear, burrs, brinelling, fatigue) is as important as part quality. Develop the proper control and monitoring functions as part of the coordinated product and workstation design. Design the workstation layout to minimize complexity, operator fatigue, repetition, and ambiguity. Embed training and other aids as part of the coordinated product and workstation design to quickly orient new operators. Design the workstation as a "flowing" process, not as a big buffer or inventory station. Tote boxes and skids should be minimized in the workstation. Develop a preventive maintenance plan as part of the coordinated design.

☑ **Design a Robust Production System.** Avoid sequential and synchronous systems that require rigid line balancing techniques. These systems are very sensitive to machine breakdowns, missing parts, tooling problems, and quality issues. The use of redundant workstations and work cells at bottleneck locations with routing flexibility can alleviate many line balancing concerns. Avoid blocking and starving of parts at workstations by designing the system as a "flow process". Look three or four product generations into the future and plan for changes, revisions, and growth due to technology and market change. Consider developing a modular (chunked) product design that facilitates anticipated change and growth. Consider using a system of standardized and rationalized fixtures and tooling early in the design project to minimize fixturing systems, tooling systems, and randomness of tools. Minimize system interruptions by including all operations that are in the process plan in the system. For example, avoid operations like heat treating and plating that are not directly under control of the production system. Use real-time shop floor tracking and status reporting. Use Statistical Process Control (SPC) to control quality at the workstation level. Always strive to minimize non-value-added processing such as storage, conveyance, load/unload operations, and post-process inspection. Also, as much as possible, use generic equipment that can be supplied by many qualified vendors. Select vendors that have a history of assuring upgradable hardware and software.

☑ **Design for Just-In-Time (JIT).** Increase quality level by installing inspection devices or procedures before, during, or after every critical process so defective items can be addressed immediately. Never add value to a defective item or send it to the next process. Have a means to stop the line if the process is in an abnormal state. Ensure quality parts from the supplier. Plan for regular preventative maintenance. Strive to reduce set-up times. Design so that much of the set-up can be done while the machine is running. Eliminate adjustments as much as possible. Eliminate unnecessary or excessive bolts and fasteners required for set-up. Reduce the number of tools required. Practice and refine the set-up procedure early in the design. Have everything available where and when required, but only in the right quantity. Minimize the need for in-process storage. Minimize work in process by making items as they are required, not just to keep the machine busy. Consider design for a pull system in which parts are pulled by the "requiring" workstation instead of being pushed by the "producing" workstation (e.g., Kanban system). Decrease the amount of raw material stored by setting up JIT deliveries from suppliers. Strive to work with the supplier to get quality parts and thus eliminate the need for incoming inspection. Design the production system layout according to parts movement and group technology, not according to machine function. Locate the entrance and exit to a group of machines close together (e.g., U-shaped layout). Use stores having a wide frontage and a small depth. Store material as close as possible to where it will be needed. Expect workers will be trained to be multi-functional. Determine standard operations and prepare a standard operations guideline for each job. In developing standard operations, strive to eliminate all wasteful moves and unnecessary handling. As the product transitions into production, review, revise, and rationalize standard operations regularly since there is always room for improvement.

Summary of Key Concepts

➢ Producibility is the resultant of good DFM.

➢ The producibility checklist can be used to guide the product design and/or it can be used as a gauge for preparing for and conducting DFM reviews.

➢ The producibility checklist can also be used to quickly come up to speed in the principles and practices of DFM.

Chapter 17
The Essence of DFM

After all is said and done, the essence of DFM comes down to five key considerations: (1) geometric layout, (2) the axioms of good design, (3) design team mindset, (4) good communication, and (5) a systematic and disciplined DFM approach. From a DFM perspective, the most important design decisions that are made over the course of a design project are those that relate to the geometric layout. Said another way, how the design is divided and arranged into individual parts sets the stage for every manufacturing and assembly decision to come. In many ways, the geometric layout single-handedly determines the fabrication description of the design. For this reason, it is extremely important that the most optimal geometric layout be identified early in the design before irreversible decisions have been made.

As a rule, *the best geometric layout is the one having the fewest easy to make and assemble parts*. Of course, like most rules, this rule must be carefully interpreted. For example, it is important that the number of parts be in the "sweet spot" that balances individual part complexity, development cost, and time to market with the theoretical minimum number of parts that is possible (see Fig. 9.1, page 136). The first geometric layout is seldom the one having the least information content. It is best to propose an initial "trial" geometric layout that maintains independence of functional requirements, analyze it considering the DFM guidelines (Chapters 6-14), and then use the insights gained to redesign for reduced information content. Importantly, the Geometric Layout Improvement Method (Chapter 12) should be used to provide a disciplined and systematic analysis approach.

Why is the geometric layout so important? Creating the geometric layout involves a complex interplay of design choices that relate to the functional requirements and constraints of the design, the spatial arrangement of parts, the configuration of individual components, the selection between standard and designed components, the selection of materials and manufacturing processes, and the assembly process. Taken together, these choices

The Essence of DFM

determine ease of assembly and lifecycle support. They also determine the number and complexity of designed parts, which in turn, influences tooling cost, tolerance stack-up, smoothness of force-flow, and numerous other considerations. As a result, geometric layout has an enormous and far-reaching effect on all aspects of the design and its method of manufacture.

Fortunately, for many assembled products, the geometric layout can be evolved and optimized through continuous improvement and cost reduction programs. But this is not ideal because it involves design effort, extra costs, and the risk of sub-optimal design, all of which can be avoided in the first place by "doing it right the first time". Therefore, to the extent possible, it is far better to spend the time and effort needed to optimize the geometric layout up front, when the product is first designed rather than relying on redesign after the fact.

The axioms of good design are the key to identifying the best geometric layout. Implementing the independence and information axioms by first creating an initial design that maintains the independence of functional requirements and applying the eliminate, simplify, and standardize where possible strategy to every aspect of the product design and production system, leads, by definition, to the best geometric layout possible given the functional requirements and constraints that must be satisfied. The information axiom underlies all aspects of DFM. Most importantly, it provides a straightforward way to measure simplicity and to therefore implement the "keep it simple" dictum.

The design axioms on their own, however, are not enough to ensure an optimize geometric layout. The lynchpin of successful DFM is a positive design team mindset. DFM must be "owned" by the team. Otherwise, it becomes just one more "hoop" the team must jump through to get the design out the door. DFM ownership is facilitated and encouraged by a company-wide standard of excellence. Such a standard helps provide the time and resources needed and creates a corporate culture and attitude that expects and nurtures effective DFM.

In addition to a positive mindset, a successful team approach depends on effective and timely communication between all stakeholder entities and activities, both internal and external, that provide input to and/or constrain the design. This means that the design team must interact efficiently with external customers and suppliers and with internal activities such as marketing, research and development, sales, purchasing, manufacturing, and service. Effective communication requires high quality communication channels. These channels should be established at the start of the project and should remain in place throughout its duration.

Downstream manufacturing and support needs impose additional requirements and constraints on the design. Without a systematic and disciplined DFM approach, there is no guarantee that these needs will be carefully considered in a timely manner. The DFM approach presented in this book is predicated on the recognition that:

- Design is the first manufacturing step.
- Every design decision, if not carefully considered, can cost extra manufacturing effort and productivity loss.
- The product design must be carefully matched to advanced flexible manufacturing, assembly, quality control, and material handling technologies to fully realize the productivity improvements promised by these technologies.

The DFM approach is simple to use, easy to remember, and is applicable to all conceivable situations. When properly implemented, it helps ensure consideration of manufacturing and support needs in all phases of the design process. Proper implementation, in turn, demands excellent communication between all stakeholders in the design and works best when the product design and production system are integrated. A product designed using the DFM approach will, by definition, be an efficient design to manufacture and assemble, either manually or automatically, and will help maximize total design value in all cases.

References

Boothroyd, G., Dewhurst, P., and Knight, W., 1994, *Product Design for Manufacture and Assembly*, Marcel Dekker, Inc., New York.

Bradyhouse, R., 1987, "The Rush for New Products Verses Quality Designs That Are Producible: Are These Objectives Compatible?" presented at the *SME Simultaneous Engineering Conference*, held June 1, 1987, Society of Mechanical Engineers, Dearborn, MI.

Dixon, J and Poli, C., 1995, *Engineering Design and Design for Manufacturing*, Conway, MA: Field Stone Publishers.

Ettlie, J. and Stoll, H., 1990, *Managing the Design-Manufacturing Process*, McGraw-Hill, NEW York.

Groover, Mikell P., 2001, *Automaton, Production Systems, and Computer-Integrated Manufacturing*, Second Edition, Prentice Hall.

Hauser, J. and Clausing, D., 1988, "The House of Quality," *Harvard Business Review*, Vol. 66, No. 3, May-June, pp. 63-73.

Shah, J. and Mantyla, M., 1995, Parametric and Feature-Based CAD/CAM, John Wiley & Sons, New York.

Stoll, H., 1999, *Product Design Methods and Practices*, Marcel Dekker, Inc., New York.

Stoll, H., 2012, *The Design for Everything Manual*, Amazon.com.

Suh, N., Bell, A., and Gossard, D., 1978, "On an Axiomatic Approach to Manufacturing and Manufacturing Systems", Journal for Engineering in Industry, vol. 100, no. 2, pp 127-130, 1978.

Suh, N. P., *Axiomatic Design: Advances and Applications*, Oxford University Press, 2001.

Taguchi, Genichi, 1993, *Taguchi on Robust Technology Development*, ASME Press, New York.

Ulrich, K. and Eppinger, S., 2000, *Product Design and Development*, McGraw-Hill, New York.

Whitney, Daniel E., *Mechanical Assemblies*, Oxford University Press, New York, 2004.

Yasuhara, M and Suh, N.P., 1980, "A Quantitative Analysis of Design Based on Axiomatic Approach", Computer Applications in Manufacturing Systems, ASME Production Engineering Div. Publication, PED-vol 2.

About the Author

Henry W. Stoll is professor emeritus of mechanical engineering at Northwestern University, Evanston, Illinois. He has extensive industrial experience in mechanical engineering design and has consulted widely with industry. Professor Stoll has taught product design, mechanical design, and manufacturing for many years. He is the author of three books on design and several book chapters on design for manufacture and has presented more than 100 on-site design for manufacture workshops to industry.

www.ingramcontent.com/pod-product-compliance
Lightning Source LLC
Chambersburg PA
CBHW071354210526